RISE ABOVE

Bridging the Divide of Hope and Fear While Taking a Deep Dive into our Nation's Concerns

By

Bernice N. King-Strong, Ph.D.

The Interconnectedness of Economics, Education, and Religion

Copyright page

Prime Publishing

Manhattan, New York

Published by Prime Publishing, April 2025

Library of Congress Control Number

Print the information available on the last page

This book is printed on acid-free paper

TABLE OF CONTENTS

PROLOGUE

This book contends that there exists an inextricable link between religion, education, and economics, which serve as foundational pillars that shape human experience, particularly in times of uncertainty. Each of these institutions provides frameworks for analyzing, understanding, and extracting meaning regarding societal stability and adaptation, both individually and collectively. This book explores how they influence personal identity and behavior, societal structures, and crisis responses by examining their connected roles. Our societal fabric's pillars shape how individuals, communities, and organizations navigate an increasingly complicated and unpredictable world.

I hope this book will describe how religion, education, and economics function as system mechanisms of resilience and control, offering guidance, opportunity, and security in times of instability. This book will provide a comprehensive understanding of how these forces act together to shape human lives and societies across different historical and cultural contexts. Through this exploration, the information will illustrate how people use belief systems, knowledge, and financial structures to manage uncertainty and shape their futures.

This book provides a deep, interdisciplinary exploration of how religion, education, and economics shape human experience, equipping readers with a nuanced understanding of these forces in both historical and modern contexts.

INTRODUCTION

Hope, Fear, and Uncertainty in Contemporary Society

In the ever-changing landscape of contemporary society, three emotions have come to define human experience: hope, fear, and uncertainty. These powerful forces guide our decisions, shape our futures, and influence our perceptions of the world around us. While these emotions may seem disparate, they are interconnected, weaving a complex fabric that defines not only the individual experience but also the collective consciousness of our society. Through the lens of these emotions, readers will better understand the challenges we face today and how we navigate the complexities of modern life.

I will offer a definition of hope better to understand a common definition of hope for this writing. Hope is often described as the belief in a better future, a sense of optimism that better days lie ahead despite our struggles today. Hope is the spark that drives us to overcome adversity, propels us forward when the road ahead seems insurmountable, and serves as a reminder that even in the darkest moments, there is the potential for change. Hope is a lifeline to hold onto when everything else seems uncertain.

In contemporary society, hope is not just a feeling but a force that drives us to make positive changes. It is essential to understand and make meaning of constant global challenges—political, environmental, or social—and people need hope to believe that positive change is possible. It is the foundation for activism, fighting injustice, and striving toward a better world. Hope compels people to act, demand better policies, fight for equality, and ensure that the next generation has a better life than those who came before. Without hope, there would be little motivation for progress. Our challenge in evolving into a community of hope for all is avoiding the dangers prevalent when personal self-interest becomes the dominant force of community members' aims.

In stark contrast to hope is the emotion of fear. Fear arises when we perceive a threat, whether real or imagined. It is a feeling that can inhibit individual action; it restricts and may prevent us from pursuing our goals. Fear exists and manifests itself in many forms: failure, fear of the unknown, fear of loss, and, most significantly, in today's world, fear of the future. It is a feeling that can restrict individual actions, often preventing people from pursuing their true desires.

The global concern over climate change has intensified in recent years. The increasing frequency of natural disasters, rising temperatures, and substantial scientific evidence pointing to irreversible damage have generated widespread apprehension in societies worldwide. For many individuals, fears regarding the future become overwhelming due to uncertainties about the planet's sustainability and the future of human civilization. The fear of extinction, of watching entire ecosystems collapse, and of children growing up in a world ravaged by environmental collapse haunts millions. It affects decisions from what we consume to how we vote and even how we raise our children.

Nevertheless, fear can also serve as a motivator. While it can immobilize, it can motivate individuals and communities to act. Fear of climate change, for instance, has led to a rise in green movements, protests, and a demand for renewable energy solutions. Fear of political instability has inspired grassroots movements advocating human rights and social justice. Fear often exposes our vulnerabilities and prompts us to take the necessary steps to safeguard ourselves and others.

It is imperative for us to understand that unchecked fear can be destructive. It can breed anxiety, prejudice, and hatred. When fear is allowed to fester and grow unchecked, it can lead to xenophobia, racism, and division within society. Fear of the "other"—whether that other is a different nationality, religion, or race—has led to the rise of nationalism, terrorism, and other harmful ideologies. Fear often finds its way into the political landscape, where it is manipulated to create fear-based policies that exacerbate division rather than foster unity.

Fear and hope are not the only emotions felt in our society. Perhaps the most pervasive emotion in contemporary society is uncertainty. The rapid pace of technological advancements, shifting political climate, and constantly changing global landscape have led to an unpredictable and unstable world. Uncertainty permeates every aspect of our lives—from the economy to international relations, from the job market to healthcare. It is a constant companion, lurking in the background, shaping our plan.

This sense of uncertainty has been amplified by the events of the past two decades. The 9/11 attacks, the fiscal crisis of 2008, the rise of global terrorism, and, more recently, the COVID-19 pandemic, have all served as reminders that the world we live in can change in an instant. These events have shattered the illusion of stability that many people once held, replacing it with the stark reality that we live in an age of volatility. How does the emotion of uncertainty manifest in our lives?

Uncertainty breeds a sense of unease, causing many to question what lies ahead. This uncertainty can feel overwhelming for young people and members of the senior population. With the future of work changing rapidly due to automation, the instability of the job market, and the ongoing challenges of climate change, it is not easy to know what the future holds. How do we plan for a future when the very ground beneath our feet feels unpredictable? Is it possible for uncertainty to serve a dual purpose?

The answer that I offer is YES. Despite its discomforting nature, uncertainty also has its role. It encourages flexibility and adaptability. The human spirit is incredibly resilient. Often, in the face of uncertainty, we find ways to cope. We cope by innovating, learning, and adapting. Uncertainty also reminds us of the transient nature of life; it forces us to embrace change rather than resist it. In a constantly evolving world, uncertainty is inevitable and essential for growth.

The Interconnectedness of Emotions of Fear, Hope, and Uncertainty

Fear, hope, and uncertainty are not isolated emotions; these emotions coexist, often in complex and intertwined ways. One does not exist without others. The interplay among them shapes how we respond to the world around us. Hope can be born from fear; often, recognizing fear motivates people to seek hope. Uncertainty, too, can spark both fear and hope. The unknown is a fertile ground for both anxiety and possibility. These emotions interact to create a dynamic force that propels society forward.

One example of the interconnectedness of three emotions was the civil rights movement in the United States. The fear of oppression, the hope for equality, and the uncertainty of the future came together to create one of the most significant social movements in modern history. In his book The Radical King, the Reverend Dr. Martin Luther King, Jr. exhorted that the problems of racism, economic exploitation, and war are all tied together. These are the interrelated triple evils. (King, ML, Jr.,2015, p. 103).[1]The fear of continued inequality and injustice spurred individuals to act, while hope for a better future fueled the movement's perseverance. The uncertainty of whether change could be achieved was a constant challenge, but it did not stop activists from pressing forward. In today's world, this interplay is evident in many areas. The fight against climate change, social justice movements, and political struggles are all driven by the push and pull of hope, fear, and uncertainty. Activists, leaders, and citizens face our time's daunting challenges. Progress can only be made through the balance of these emotions. The hope for a better world must be nurtured, and we must acknowledge the fears that drive us to action. Therefore, uncertainty must not be seen as a barrier but as a challenge that calls for innovation, adaptation, and resilience.

We must embrace hope, fear, and uncertainty as fundamental aspects of human experience and see them as crucial drivers of progress in contemporary society. While hope gives us direction and purpose, fear pushes us to act, and uncertainty encourages flexibility and growth. Together, these emotions shape the decisions we make, the paths we take, and the future we hope to build.

We live in a world marked by change and instability; it is essential that we learn to navigate these emotions with awareness and intention. By embracing hope, confronting fear, and adapting to uncertainty, we can continue to evolve and strive toward a more just, sustainable, and equitable world. The challenges we face today are

---

[1] King, Martin Luther, Jr. "The Radical King", p.103, Beacon Press

daunting, but they also present opportunities for growth and transformation. If we hold on to hope, confront our fears, and face uncertainty with courage, we can move forward to a brighter future than we can imagine.

Chapter 1

THE ROOTS OF HOPE

A Historical, Psychological, and Sociological Perspective

Hope is an expectation or anticipation regarding a future outcome, whether it is something desired to occur, believed to be true, or something not wanted to happen or not believed to be true. Hope is a fundamental human emotion influencing thoughts, behaviors, and societal structures. It serves as a driving force in personal aspirations, social movements, and the progress of civilizations. Hope can be defined as an optimistic state of mind based on an expectation of positive outcomes related to events and circumstances in one's life or the world at large. This introductory chapter explores the nature of hope by delving into historical perspectives and the psychological and sociological aspects of this powerful emotion.

Hope is an experience that encompasses both rational and emotional elements, allowing individuals to envision a future better than their current circumstances. It frequently emerges during challenging times, protecting against despair and encouraging perseverance despite obstacles. The concept of hope has been extensively discussed by philosophers, theologians, and scientists, with interpretations ranging from existential necessity to simple aspiration. From a philosophical standpoint, hope has been addressed as a theological virtue, not a moral imperative. (Bobier, 2017, Albo,1475). Ancient philosophers such as Aristotle and Plato viewed hope as an essential component of human experience, often linked to faith and the pursuit of a good life. In contrast, some existentialist thinkers, such as Friedrich Nietzsche, warned against false hope, suggesting that it might be an illusion preventing individuals from confronting harsh realities (Nietzsche, 2012). Despite these differing viewpoints, hope remains an enduring and universal aspect of human existence.

Historically, hope has played a crucial role throughout the past, particularly during times of crisis and transformation. Religious traditions across cultures have often emphasized hope as a guiding principle. Christianity, for example, presents hope as one of the three theological virtues, alongside faith and charity. Biblical hope represents a confident expectation rooted in faith and trust in God. Rather than a fleeting desire or wish, it embodies assurance and promises of future fulfillment. This hope is not passive; it calls for active engagement in life, encouraging you to persevere even through hardships. Biblical scripture frequently references hope as a source of strength and resilience, particularly in challenging times (Romans 15:13, KJV).

Similarly, hope has been a catalyst for social and political change. Movements such as the abolition of slavery, civil rights struggles, and feminist activism have all been fueled by hope. Leaders like Martin Luther King Jr. emphasized hope as a means of envisioning and working toward a just society. His famous "I Have a Dream" speech is a testament to the power of hope in inspiring collective action (1963). Hope has been a crucial survival mechanism even in the face of extreme adversity, and during historical periods such as the Holocaust, prisoners in concentration camps often clung to hope as a means of psychological endurance (Frankl, 1946). Psychologist Viktor Frankl, a Holocaust survivor, wrote in his book "Man's Search for Meaning" that hope and a sense of purpose were critical factors in surviving unimaginable suffering (Frankl, 1946). Hope plays a vital role in mental health and well-being. Positive psychology, led by Martin Seligman (1998), examines how hope boosts resilience, reduces anxiety, and improves life satisfaction. Research shows hopeful individuals cope with stress, recover from illness faster, and achieve their goals more effectively.

The "hope theory" concept developed by psychologist C.R. Snyder (1991) suggests that hope consists of two primary components: agency and pathways. Agency refers to the motivation and belief that one can achieve desired outcomes, while pathways involve identifying strategies to reach those goals. This theory underscores the active nature of hope, distinguishing it from passive dreaming. Additionally, hope is intricately linked to mental health. Research indicates that individuals with elevated levels of hope tend to have lower rates of depression and anxiety (Snyder, 1991). Hope can be a protective factor, providing individuals with the psychological resources to navigate life's challenges. Sociological aspects of hope extend beyond the individual and play a vital role in shaping societies. Sociologists examine hope as a collective phenomenon influencing group behavior, social movements, and cultural development (Scribano,2026). In times of economic downturn, political instability, or natural disasters, hope is a unifying force that mobilizes communities toward recovery and progress. Examples of this kind of mobilization are social movements, such as the Civil Rights Movement in the United States and the Arab Spring wave of pro-democracy protests and uprisings that took place in the Middle East and North Africa beginning in 2010 and 2011, challenging some of the region's entrenched authoritarian regimes demonstrate how collective hope can drive societal transformation. When people believe in the possibility of change, they are more likely to act and work toward a shared future vision. Sociologists also study the role of hope in marginalized communities, where it functions as a coping mechanism and a means of challenging systemic oppression.

Additionally, hope is a key element in education and social development. Educators and policymakers recognize the importance of fostering hope in young people, as it enhances motivation and academic achievement. Programs aimed at empowering disadvantaged youth often emphasize the cultivation of hope to break cycles of poverty and inequality (Bhadra,2023). Hope is powerful and multifaceted. It is a force that influences individuals and societies alike. Hope has driven human progress from historical movements to psychological well-being and sociological transformations. While perspectives on hope vary, its role in providing resilience, motivation, and collective action remains undeniable. Individuals and communities can strive toward a future filled with possibility and positive change by understanding and nurturing hope.

Chapter 2

THE ANATOMY OF FEAR

Fear is a fundamental human emotion that has played a crucial role in survival and evolution. It is a psychological and physiological response to perceived threats or dangers, preparing individuals to either confront or avoid the source of their distress (Cannon,1932). While fear is often viewed negatively, it can be essential in safeguarding individuals from harm. Understanding fear requires an exploration of its origins, diverse types, impact on human behavior and decision-making, and how societal forces manipulate it. The roots of fear lie in human biology and psychology. From an evolutionary perspective, fear developed as a mechanism to alert individuals to potential threats, increasing their chances of survival (Cannon, 1932). The brain's amygdala plays a vital role in processing fear, activating the body's fight-or-flight response. This response results in physiological changes such as an increased heart rate, rapid breathing, and heightened sensory perception Cannon, 1932.

Psychologically, fear can also be learned through firsthand experiences or cultural conditioning. As demonstrated in Pavlovian experiments, classical conditioning explains how individuals associate certain stimuli with fear responses. Similarly, observational learning, where people witness others experiencing fear, can reinforce and shape their fears. Additionally, genetic predispositions can influence susceptibility to fear-related disorders, highlighting the interplay between biology and environment in fear development. According to (Frankl, 1955 Williamson, 1992), fear manifests in various forms, ranging from immediate, instinctual responses to long-term, deeply ingrained anxieties. These can be categorized into the following types:

There are several types of Fear:

1. **Innate Fears**: These are hardwired fears humans are born with, such as the fear of loud noises or sudden movements. They are primarily linked to survival instincts.

2. **Conditioned Fears**: These arise from individual experiences or societal influences. A person who has been in a car accident may develop a fear of driving, while cultural narratives can instill fears of certain people or situations.

3. **Phobias**: Phobias are intense, irrational fears of specific objects, situations, or concepts. Examples include arachnophobia (fear of spiders) and agoraphobia (fear of open spaces). These fears can significantly impact a person's daily life.

4. **Existential Fears**: These fears pertain to broader human concerns, such as the fear of death, meaninglessness, or loss of identity. Such fears often manifest in anxiety and philosophical contemplation.

5. **Social Fears**: Fear of rejection, embarrassment, or failure falls into this category. Social fears can influence how individuals interact with others and make decisions in professional and personal settings.

According to (Guy-Evans,2025), fear profoundly influences human behavior, often overriding rational thought and guiding decision-making processes. The following are some of the fundamental ways in which fear shapes human actions:

1. **Fight-or-Flight Response**: When confronted with immediate threats, the brain triggers either an aggressive response (fight) or an attempt to escape (flight). This instinctual reaction is crucial for survival but can also lead to impulsive and irrational behavior.

2. **Paralysis and Avoidance**: In some cases, fear can cause individuals to freeze or avoid situations altogether. It is clear in cases of social anxiety, where fear of judgment may prevent people from engaging in social interactions.

3. **Influence on Risk Assessment**: Fear can distort risk perception, making individuals overestimate dangers and

can lead to overly cautious behavior or, conversely, risky behavior if fear is ignored or suppressed.

4. **Emotional and Cognitive Biases**: Fear can lead to cognitive biases such as confirmation bias, where individuals seek information that supports their fears while ignoring contradictory evidence. It can perpetuate irrational fears and hinder logical decision-making.

5. **Long-Term Psychological Effects**: Chronic fear or anxiety can contribute to mental health disorders such as post-traumatic stress disorder (PTSD), generalized anxiety disorder, and depression, significantly affecting a person's quality of life.

Fear is not only a personal experience but also a powerful tool that societal forces use to influence public perception and behavior. (Cannon, 1932) Governments, media, corporations, and interest groups often exploit fear for various motives, including political control, economic gain, and social conformity.

Examples of how the emotion of fear is used include:

1. **Political Fearmongering**: Leaders and governments have historically used fear to combine power, justify wars, and suppress dissent. By emphasizing real or exaggerated threats, political entities can persuade the public to support policies they might otherwise resist, such as surveillance programs or military interventions.

2. **Media Sensationalism**: News organizations often highlight fear-inducing stories, as fear attracts viewership and engagement. Constant exposure to alarming headlines can create a heightened sense of insecurity and anxiety in society, shaping public attitudes and behaviors.

3. **Corporate Marketing Tactics**: Fear is a common strategy in advertising, where companies capitalize on consumer insecurities to sell products. From anti-aging creams to security systems, businesses create a sense of necessity by instilling fear of negative outcomes.

4. **Religious and Cultural Conditioning**: Religious and cultural institutions have used fear of divine punishment,

social exclusion, or moral failure to regulate behavior and control communities.

5. **Fear-Based Social Movements**: While fear can be a catalyst for social change, extremist groups can also exploit it to recruit members or justify radical actions. Fear of outsiders, economic instability, or ideological threats can be weaponized to mobilize support for divisive agendas.

Fear is a complex and deeply ingrained aspect of human psychology, originating from both biological imperatives and learned experiences. While it serves a crucial role in self-preservation, it also has significant implications for behavior, decision-making, and societal dynamics. Understanding the several types of fear and recognizing its manipulation by external forces can empower individuals to navigate fear more rationally and critically. By fostering awareness and emotional resilience, people can mitigate the negative effects of fear and make more informed decisions in their personal and societal lives.

Chapter 3

THE STATE OF UNCERTAINTY

Uncertainty is an inherent aspect of life. Change remains the only constant, whether we welcome it or not. Frequently, we encounter unforeseen challenges and circumstances in a perpetually evolving world. It is during these uncertain times that our resilience and inner strength are truly evaluated. Rather than resisting change, we should embrace uncertainty and leverage it as an opportunity for growth and self-development. Throughout history, wise men and women have offered words of wisdom and inspiration to help us navigate the uncertain waters of life. From philosophers and poets to leaders and visionaries, their powerful quotes have stood the test of time and continue to resonate with us today. These quotes remind us that uncertainty is not something to be feared but rather a powerful force that can push us to new heights of resilience and courage:

"Life is a series of natural and spontaneous changes. Don't resist them; that only creates sorrow. Let reality be reality. Let things flow naturally forward in whatever way they like." – Lao Tzu

"The only way to make sense out of change is to plunge into it, move with it, and join the dance." – Alan Watts

"In the middle of difficulty lies opportunity." – Albert Einstein

"There is no education like adversity." – Benjamin Disraeli

As an inevitable aspect of life, uncertainty is a major topic of discussion in Christian biblical scriptures (James 4: 13-15, Matthew 6:34, and Proverbs 3: 5-6, KJV). It encompasses situations where the future is unpredictable, where outcomes are unknown, and where ambiguity prevails (Kierkegaard, S., 1846). Nietzsche, F., 1886; and Brown, B.2012). From personal decisions to global events, uncertainty profoundly influences individuals and societies. This chapter explores the manifestations of uncertainty, its effects on both personal and collective levels, and the strategies people and societies adopt to manage and mitigate its impact.

Uncertainty manifests in various forms, affecting individuals, businesses, governments, and entire communities. It can arise in multiple contexts, including economic instability, health crises, environmental changes, and political upheavals.

1. **Personal Uncertainty:** On an individual level, uncertainty appears in career choices, relationships, and health concerns. People often face dilemmas about their future, whether in pursuing education, selecting a job, or navigating personal relationships. Health uncertainties, such as an unexpected diagnosis or concerns about aging, further add to personal anxieties.

2. **Economic and Financial Uncertainty:** Economic instability, market fluctuations, and unemployment create uncertainty for individuals and businesses. Sudden financial crises or job losses can disrupt lives and force individuals to make difficult adjustments.

3. **Political and Social Uncertainty:** Political instability, wars, and policy changes generate societal uncertainty. Citizens may struggle with unpredictable governance, loss of rights, or safety concerns in politically unstable regions.

4. **Environmental and Climate Uncertainty:** The unpredictability of climate change, natural disasters, and environmental degradation presents another form of uncertainty. Individuals and societies must prepare for extreme weather events, resource scarcity, and long-term ecological shifts.

Effects of Uncertainty on Individuals and Societies

Uncertainty is a fundamental and unavoidable aspect of life, profoundly shaping individuals and societies. The nature of uncertainty can be explored through its psychological, social, and cultural effects. For individuals, uncertainty often triggers a mix of emotions—curiosity, anxiety, hope, or fear. It can compel people to seek security, predictability, or control, influencing decision-making and behavior. For example, in the face of uncertainty, some may cling to established routines or beliefs, while others may embrace change and exploration. This duality shows how uncertainty can either paralyze or empower, depending on how it's perceived and managed. On a societal level, uncertainty is both a challenge and a catalyst. During times of uncertainty, such as economic instability, political upheaval, or global crises like pandemics, societies may experience heightened tensions, polarization, or collaboration. Uncertainty often drives innovation and resilience, inspiring communities to adapt and create solutions. However, it can also amplify misinformation, fear, or inequality if not carefully navigated[2].

Culturally, societies develop narratives, rituals, and systems—like religion, science, or philosophy—to address uncertainty, offering meaning and frameworks for coping with the unknown. These systems help mitigate the collective anxiety that uncertainty can bring and inspire progress and transformation. So, uncertainty is both a disruptor and a driver, shaping human experiences on every level. It's a force that can either hinder or propel growth, depending on how it's understood and approached.

Uncertainty has significant psychological, emotional, and social effects. It influences decision-making, mental health, and societal stability.

1. **Psychological and Emotional Effects:**
   - Increased stress and anxiety due to fear of the unknown.
   - Difficulty in making decisions, leading to procrastination or avoidance.
   - Depression or helplessness when uncertainty feels overwhelming or unmanageable.
2. **Impact on Decision-Making:**
   - People may resort to risk-averse behaviors, avoiding challenges or change.

---

[2] Seligman, M & Csikszentmihalyi, M. (2000). *Positive Psychology: An Introduction. American Psychologist.*

- Over-reliance on heuristics or impulsive decision-making as a way to cope.
- Fear of making the wrong choice may lead to indecision and inaction.

3. **Social and Economic Consequences:**
   - Uncertainty can lead to distrust in institutions, governments, or businesses.
   - Economic instability can result in decreased investments and spending, slowing growth.
   - Societal fragmentation may occur as groups react differently to uncertainty, leading to polarization or conflict.

Coping Mechanisms and Strategies

Despite its pervasive nature, uncertainty can be managed through effective coping strategies. Individuals and societies adopt different approaches to mitigate uncertainty's negative effects and foster resilience.

1. **Psychological Coping Mechanisms:**
   - **Mindfulness and Acceptance:** Accepting that uncertainty is a natural part of life reduces anxiety and fosters mental well-being.
   - **Resilience Building:** Developing adaptability and emotional strength helps individuals face uncertainty without excessive distress.
   - **Cognitive Reframing:** Shifting perspectives to see uncertainty as an opportunity for growth rather than a threat.
2. **Practical Strategies for Individuals:**
   - **Information Gathering:** Seeking reliable information reduces uncertainty in decision-making.
   - **Financial Planning:** Having an emergency fund or diversified investments minimizes economic uncertainty.

- **Developing Contingency Plans:** Preparing for multiple scenarios helps mitigate the impact of unexpected changes.

3. **Societal and Institutional Approaches:**
    - **Government Policies and Safety Nets:** Social security programs, unemployment benefits, and healthcare access reduce the effects of economic and health uncertainties.
    - **Scientific Research and Technological Innovations:** Advances in science and technology help predict and manage uncertainties, such as climate forecasting and medical breakthroughs.
    - **Education and Awareness:** Promoting critical thinking and adaptability through education empowers individuals to manage uncertainty effectively.

The renowned philosopher Lao Tzu (6th. Century BCE) taught that "Life is a series of natural and spontaneous changes. Don't resist them; that only creates sorrow. Let reality be reality. Let things flow naturally forward in whatever way they like." As an element of one's natural changes throughout life, uncertainty is a fundamental aspect of life, shaping decisions, emotions, and societal dynamics. While its effects can be challenging, individuals and societies have developed various mechanisms to cope with its unpredictability. By fostering resilience, strategically planning, and embracing adaptability, individuals and communities can confidently navigate uncertainty. Rather than viewing uncertainty as a threat, embracing it as an opportunity for growth and learning allows for a more balanced and prepared approach to life's challenges.

Chapter 4

EDUCATIONAL ASPIRATIONS AND REALITIES

Current students—primary, secondary, or collegiate—face significant anxiety due to performance and future prospects (Talley, 2024). This fear affects their mental well-being, academics, and overall development. While academic pressure is longstanding and can be motivational, rising competition, high expectations, and a focus on grades and careers have exacerbated insecurity. This chapter examines the sources and effects of performance-related anxiety, its impact on mental health and achievements, and ways to mitigate these fears in schools and other educational institutions.

Some Sources of Fear in Educational Settings

High Academic Expectations often cause students to feel anxious due to the pressure to achieve top grades, excel in extracurricular activities, and be on the honor roll. This expectation for success can be overwhelming and lead to significant stress from parents, teachers, peers, and the students themselves.

Parents often have high expectations for their children's academic performance, aiming to secure better future opportunities. These expectations can be expressed as pressure, with parents stressing the importance of good grades and higher education. Teachers can also contribute to students' stress by setting ambitious standards for assignments and exams. Although these expectations are meant to motivate students, they can lead to anxiety if the student has difficulty meeting them (Haider, 2022).

Concerns About Future Prospects Besides immediate academic performance, students often think about their future opportunities. Concerns about obtaining a favorable job, gaining admission to a prestigious college, or achieving their career goals can affect students. The increasing cost of higher education and the competitive job market further heighten these concerns. With society placing importance on academic success as a predictor of future achievement, students may feel that their future depends significantly on their classroom performance.

In the age of social media, students are exposed to highly curated portrayals of success and achievement, which can intensify feelings of inadequacy and fear of missing out. When students compare themselves to peers who excel in academics or extracurriculars, this comparison can further contribute to performance anxiety, as students may feel that they are falling behind in a difficult race to win.

Fear of Failure is a notable source of anxiety for many students. This fear often exacerbates societal perceptions that equate failure with personal inadequacy. In educational contexts, this apprehension can lead to perfectionism, procrastination, or the avoidance of challenging tasks. Students may become so inhibited by the fear of failing that they encounter difficulties completing assignments, performing in examinations, or engaging in risk-taking behaviors essential for learning. Consequently, this often leads to reduced academic performance, thereby reinforcing their sense of failure and associated anxiety.

Many students fear failure, which can lead to anxiety and avoidance, hindering their academic growth. This performance anxiety can have significant negative impacts on their education and future career aspirations.

The Consequences of Performance Anxiety

Mental Health Struggles resulting from fear of poor performance and uncertain futures can harm students' mental health, leading to chronic anxiety, depression, and stress (Bernal-Morales et al., 2015). Pressure to excel can cause burnout, impairing physical, emotional, and mental functioning.

Academic pressure can trigger severe mental health issues like anxiety disorders or depression. This may lead to a cycle where fear affects academic performance, resulting in more anxiety. Performance anxiety can escalate to self-harm or suicidal thoughts, underscoring the need to address this within educational settings.

Academic Underachievement Fear of failure can lead to poor academic performance. Performance anxiety may cause students to procrastinate, manage time poorly, and avoid challenging tasks, hindering their learning. In extreme cases, they might withdraw from academics, believing success is unattainable (Jiang et al., 2022).

Performance anxiety can cause students to see education as just a path to grades, college, or careers rather than a chance for personal growth. This narrow view can reduce their engagement and enthusiasm for learning. As students become increasingly preoccupied with their academic performance, they may withdraw from social interactions. This isolation can result from their fear of not measuring up to their peers or believing their worth is solely tied to their academic achievements. Social interactions are crucial for emotional development, and when students cut themselves off from their support networks, they may feel even more isolated and anxious.

Competitiveness in academic settings can impact feelings of isolation. Students may perceive continuous competition with their peers, which can affect the sense of community and result in emotions such as jealousy, inadequacy, and resentment.

Strategies to Alleviate Fear in Educational Settings

Psychologist Carol Dweck (2016) promotes a Growth Mindset as an essential approach to student success. The concept of a growth mindset suggests that intelligence and abilities can be developed through effort and perseverance. This approach prioritizes learning processes over outcomes such as grades, encouraging students to view challenges as opportunities for improvement rather than threats to self-worth. Recognizing that mistakes are a part of the learning process can help students understand their potential.

Emotional and Psychological Support is increasingly important in educational institutions. Schools and universities can play a vital role in addressing students' fears and anxieties by offering counseling services, workshops on stress management, and peer support networks to help students cope with the pressures they face. Creating an environment where students feel safe expressing their concerns and seeking help is crucial for their well-being.

In addition, teachers and administrators can be trained to recognize signs of anxiety and depression in students, ensuring that they can offer proper support or referrals to mental health professionals. By normalizing discussions about mental health and reducing the stigma associated with seeking help, educational institutions can create a more supportive environment for all students.

Encouraging Balanced Expectations is increasingly important for parents and teachers who accept their role in fostering student success, health, and well-being. They must be mindful of the expectations they place on students. While it is important to challenge students and encourage them to strive for excellence, it is equally important to recognize their individual strengths, needs, and limitations possessed by their students. Setting realistic and attainable goals while simultaneously providing encouragement and constructive feedback can help reduce the fear of failure and foster a sense of accomplishment.

Parents should be encouraged to value their children's well-being above their academic achievements. By focusing on the importance of effort, resilience, and personal growth rather than solely on grades and outcomes, parents can create a more supportive and less stressful environment for their children.

Students' fear about performance and future prospects within educational settings is a complex issue with far-reaching implications for their mental health, academic success, and personal growth. By understanding the sources of this fear and its potential consequences, educators, parents, and students can work together to create a more supportive and less anxiety-inducing environment. Fostering a growth mindset, providing emotional support, and promoting balanced expectations can help alleviate students' fears and encourage a healthier approach to education. Education should be about more than just grades and outcomes—it should be about nurturing curiosity, resilience, and a lifelong love of learning.**Psychological Dynamics within Educational Settings**

Education has long been regarded as one of the most potent forces in shaping individuals, societies, and civilizations (Dewey, 2011). The traditional view of education is that it is a means of acquiring knowledge, developing skills, and preparing individuals for productive societal roles. However, the psychological dynamics within educational settings reveal that education functions far more complexly, exerting both positive and negative influences on the individuals involved. While education can offer hope, empowerment, and opportunities for personal growth, it can also evoke fear, anxiety, and self-doubt. Hope and fear are deeply intertwined in educational settings, impacting students, teachers, and parents alike. This section will examine how education can inspire hope and induce fear. Understanding these psychological dynamics can foster positive emotional development and mitigate adverse effects in educational environments. This leads us to consider the role of education in our lives and the influence of hope on learning.

The Role of Hope in Education

Hope is a powerful psychological force that can propel individuals forward, even in the face of challenges. Hope's role in educational settings manifests in various forms, from students aspiring to achieve academic success to teachers believing in their students' potential to succeed. Hope catalyzes motivation and perseverance, providing individuals with the belief that their efforts will lead to meaningful outcomes.

Hope also serves as a motivational force for parents, students, and teachers. For students, education represents a pathway to future success. The hope of achieving academic goals, gaining acceptance into higher education institutions, and securing better job prospects often drives students to push through the difficulties of their studies. In the classroom, students are often motivated by the belief that their hard work will be rewarded through grades, recognition, or personal growth. This hope helps them overcome obstacles and persist when faced with challenges such as difficult assignments, exams, or interpersonal conflicts. The hope of achieving academic success can also give students a sense of purpose and direction, giving them something to strive toward in their lives.

The Teacher's Role in Fostering Hope

Teachers play a critical role in nurturing hope within their students. By creating supportive and encouraging classroom environments, teachers can instill in students a sense of possibility and potential (Hawthorne, 2024). Positive reinforcement, constructive feedback, and empathetic understanding from teachers can help students build self-confidence and keep a hopeful outlook on their academic journeys. Furthermore, when teachers express belief in their student's abilities, they contribute to a positive feedback loop where students internalize this belief and are more likely to put forth the effort needed to succeed.

Educational institutions that prioritize student well-being and provide resources for emotional and psychological support can further amplify hope. Counseling services, mentorship programs, and peer support networks can serve as sources of encouragement and help students navigate the emotional challenges often accompanying the educational process. When students feel they have the necessary support to achieve their goals, they are more likely to maintain hope and continue pursuing their educational aspirations.

The Role of Fear in Education

While hope is a driving force in education, fear is the darker side of the educational experience. Fear can manifest in various forms, from fear of failure to fear of judgment and social exclusion. These fears can significantly impact students' emotional well-being, leading to anxiety, stress, and diminished self-esteem. It is essential to recognize that fear in educational settings is not an inherent or unavoidable aspect of learning; rather, it is often a product of how education is structured and delivered and the psychological pressures placed on students.

Fear of Failure and Academic Pressure

One of the most prominent sources of fear in educational settings is the fear of failure. Many students experience anxiety about their ability to meet academic expectations and fear the consequences of failing. This fear can be worsened by societal pressures emphasizing achievement and success, creating a climate where failure is stigmatized and viewed as a personal shortcoming. The fear of failure can cause students to avoid taking risks, engage in avoidance behaviors, and experience chronic stress. It can also lead to a fixed mindset, where students believe their intelligence and abilities are static and cannot be improved through effort.

The pressure to perform academically can be overwhelming, particularly for students in competitive environments. Constant comparison to peers, pressure to achieve high grades, and emphasis on standardized testing can create a climate of fear where students feel that their academic performance determines their worth. This fear of not measuring up can undermine students' self-esteem and hinder their ability to engage fully with the learning process.

Fear of Social Judgment and Exclusion

In addition to academic pressures, fear in education can also stem from social dynamics. Students often worry about how they are perceived by their peers and teachers, fearing judgment, ridicule, or exclusion. Social anxiety is common in educational settings, particularly during formative years when students are developing their identities and seeking social acceptance (Boyle, 2021). The fear of being judged for appearance, abilities, or social status can make students self-conscious, anxious, and withdrawn. For some, this fear of social rejection can even prevent them from participating in class discussions or extracurricular activities, limiting their opportunities for personal growth and development.

Teachers and school administrators also contribute to the social climate of fear in educational settings. In some cases, an emphasis on discipline, compliance, and conformity to community expectations and practices can create a rigid environment where students feel that they must adhere to strict norms to be accepted. In such environments, students who deviate from the norm may experience fear of ostracization or punishment. This fear can contribute to a negative cycle where students become disengaged, disconnected, and unable to embrace the learning experience fully.

Balancing Hope and Fear

While hope and fear are present in educational settings, balancing these two psychological forces is the key to creating a positive and productive learning environment. Educational institutions must strive to cultivate an atmosphere where hope is nurtured while fear is minimized. This balance can be achieved through supportive teaching practices, emotional regulation, and a commitment to fostering positive relationships within the school community.

Balancing hope and fear is a fascinating psychological topic, often explored through various theoretical perspectives. Here are some key approaches and expert proponents:

1. **Hope Theory by C.R. Snyder**:
   - Snyder's cognitive model of hope emphasizes the ability to set goals, create pathways to achieve them and maintain the motivation to follow those pathways. This theory suggests that hope is critical in overcoming fear and achieving well-being.
2. **Polar Perspective on Hope and Fear**:
   - This perspective views hope and fear as interconnected emotions, often existing simultaneously. It suggests that hope can counterbalance fear, helping individuals navigate uncertainty and challenges.
3. **Positive Psychology and Hope**:
   - Positive psychology, as discussed in works like *The Oxford Handbook of Positive Psychology*, highlights the role of hope in fostering resilience and emotional well-being. It also explores how hope can mitigate the paralyzing effects of fear.

4. **Evolutionary Perspective**
    - Some theories propose that both hope and fear have evolutionary purposes. Fear drives individuals away from danger, while hope motivates them toward positive outcomes, creating a dynamic balance.

Educational leaders play a crucial role in fostering environments where students can balance hope and fear effectively. Here are some strategies supported by expert insights:

1. **Integrating Hope Education**:
    - Research by O'Shea et al. emphasizes the importance of incorporating hope education into the curriculum. It involves teaching students to set goals, develop strategies to achieve them and sustain motivation, which can counterbalance fear.
2. **Promoting Emotional Intelligence**:
    - Berkovich and Eyal highlight the significance of emotional capabilities in school leadership. Leaders can support students by creating emotionally safe spaces and teaching them to manage their emotions effectively.
3. **Critical Pedagogy of Hope**:
    - Inspired by Paulo Freire's concept of the pedagogy of hope, Douglas Bourn suggests that education should address real-life challenges and promote optimism about positive change. This approach can help students navigate fears about global issues.
4. **Supporting Teachers**:
    - Leaders can empower teachers to integrate social-emotional learning into their lessons, as suggested by Grissom et al. This helps students develop resilience and a balanced perspective on hope and fear.

These strategies enhance students' emotional well-being and prepare them to face life's uncertainties confidently. These perspectives provide a rich framework for understanding how hope and fear interact and influence human behavior.

Encouraging Growth Mindset

Promoting a growth mindset is one of the most effective ways to reduce fear and enhance hope in educational settings. Developed by psychologist Carol Dweck, the growth mindset is the belief that abilities and intelligence can be developed through effort and perseverance. When students adopt a growth mindset, they are more likely to view challenges as opportunities for growth rather than threats to their self-worth. Teachers can foster a growth mindset by praising effort rather than innate ability, encouraging students to embrace mistakes as part of the learning process, and emphasizing the value of persistence and resilience.

Reducing the Stigma of Failure

To address the fear of failure, educators can work to reduce the stigma surrounding mistakes and setbacks. By normalizing failure as a natural and essential part of the learning process, educators can help students reframe their thinking and reduce the psychological burden of academic challenges. Teachers can share their own experiences of failure and highlight the lessons learned from mistakes. Creating a classroom culture where failure is not punished but seen as an opportunity to gain experience can empower students to take risks, be creative, and engage more fully with the material.

Building a Supportive School Community

Finally, educational institutions must prioritize creating inclusive and supportive school communities to counteract the fear of social judgment and exclusion. Programs that promote diversity, equity, and inclusion can help students feel valued and accepted, regardless of their backgrounds or differences. Anti-bullying initiatives, peer mentorship programs, and open channels of communication between students and teachers can foster a sense of belonging and reduce the fear of social rejection. When students feel emotionally safe in their school environment, they are more likely to thrive academically and emotionally.

The psychological dynamics within educational settings are complex and multifaceted. In its ideal form, education is a source of hope, offering individuals the opportunity to grow, learn, and achieve their goals. However, the pressures associated with academic performance, social judgment, and conformity can also give rise to fear, anxiety, and self-doubt. The challenge for educators and educational institutions is to balance these two psychological forces, creating environments that foster hope while mitigating the adverse effects of fear. By promoting a growth mindset, reducing the stigma of failure, and building supportive communities, we can help students navigate the emotional challenges of education and create a more positive, fulfilling educational experience for all. Through these efforts, education can become a force for empowerment, offering hope for a brighter future while alleviating the fears often accompanying the learning journey.

Changing Educational Landscapes

The Power of Online Learning: Democratizing Education for All

Education has long been considered the cornerstone of opportunity and success. Throughout history, access to quality education has been restricted by numerous factors, including geographical location, socioeconomic status, and institutional barriers. However, the advent of the internet and the proliferation of online learning platforms have made significant strides toward overcoming these challenges, offering a more inclusive and accessible approach to education. This essay will explore how online learning democratizes education, providing opportunities for individuals who may not have access to traditional institutions. It will address how online learning makes education more accessible, flexible, and affordable, as well as its transformative impact on global education systems.

The Accessibility of Online Learning

One of the most significant advantages of online learning is its ability to reach individuals in remote or underserved areas. Traditional education often requires students to attend physical classrooms, which can be a barrier for those living in rural or impoverished regions. Access to education can be limited or even nonexistent for students in developing countries or those living in conflict zones. Online learning eliminates the need for physical attendance, allowing students to learn from anywhere with an internet connection. This accessibility opens doors to opportunities for individuals who would otherwise be excluded from educational opportunities.

For example, platforms like Coursera, edX, and Khan Academy offer a wide range of courses that anyone with an internet connection can access. These platforms provide educational content from prestigious universities and institutions, giving students from all backgrounds access to high-quality instruction. In countries where educational infrastructure is lacking, online learning offers an alternative solution, allowing individuals to learn at their own pace and on their own terms. In this way, online learning breaks down the geographical barriers that have historically limited access to education.

Furthermore, online learning enables access to many subjects and specialties that may not be available in local schools or universities. Students can choose from courses in fields such as technology, business, healthcare, and the arts, which may be in high demand but underrepresented in certain regions. Access to specialized knowledge and skills through online learning empowers students to pursue their interests and career goals, regardless of location.

This opportunity for flexibility in learning provides accessibility. Online learning offers a level of flexibility that traditional institutions often cannot match. Many people face challenges balancing education with work, family obligations, or other commitments. Attending traditional classes may be difficult for working adults or parents due to rigid schedules and location constraints. Online learning platforms provide flexibility to learn at one's own pace, on one's schedule, and from the comfort of home.

This flexibility allows individuals to continue their education without sacrificing their other responsibilities.

This is particularly beneficial for nontraditional students—individuals who may not fit the typical mold of a full-time, on-campus student. Online education allows them to study part-time, take breaks when needed, and access resources conveniently. This can be especially important for individuals reentering the workforce, seeking to upgrade their skills, or transitioning to new careers. For example, an adult with a full-time job may take an online course in the evenings or on weekends, enabling them to improve their qualifications without disrupting their daily routine.

Also, online learning's asynchronous nature allows students from different time zones or countries to participate in the same course. This creates a truly global classroom where students can engage with peers worldwide, broadening their perspectives and fostering a more inclusive learning environment. The flexibility and accessibility of online learning also enable individuals with disabilities or health concerns to participate in education in a way that might not be possible in a traditional classroom setting. Online learning can provide a more comfortable and accommodating environment for students with physical or mental health challenges.

Another key aspect of how online learning democratizes education is its potential to reduce the financial barriers associated with traditional education. The cost of attending a traditional college or university has steadily risen in recent years, with tuition fees, textbooks, and room and board becoming increasingly unaffordable for many students. For individuals from low-income families, the cost of higher education can be prohibitive, leaving them with limited options for advancing their education.

Online learning presents a more affordable alternative. Many online courses are free or offered at a significantly lower cost than traditional university programs. Websites like Coursera and edX offer free courses in various subjects, with the choice of purchasing a certificate upon completion for those who wish to highlight their accomplishments. Even when online courses do require payment, the cost is often much lower than attending a physical institution.

This affordability makes education accessible to a broader range of individuals, regardless of their financial situation.

In addition to reducing tuition costs, online learning eliminates many additional expenses associated with traditional education. Students no longer need to pay for transportation, housing, or textbooks, which can add up quickly. For learners in developing countries or those facing financial hardship, the affordability of online learning can be a notable change. It allows them to pursue education without incurring crippling debt or sacrificing necessities.

The rise of open educational resources (OER) further enhances the affordability of online learning. OERs are freely accessible, openly licensed educational materials that can be used for teaching, learning, and research. These resources include textbooks, videos, lecture notes, and other learning materials available online for free or at a minimal cost. The widespread availability of OER has the potential to make education even more affordable and accessible, particularly for students in low-income regions or those who may not have access to expensive textbooks.

Online learning has impacted educational systems globally. It can potentially transform education systems worldwide by making high-quality education more accessible and equitable. In many countries, educational resources are scarce, and the quality of education can vary significantly between schools. Online learning can help level the playing field by providing students access to top-tier educational content, regardless of where they live or attend school.

Furthermore, online learning promotes lifelong learning, encouraging individuals to continue their education beyond traditional schooling. In a rapidly changing world, acquiring new skills and knowledge is crucial for adapting to technological advancements and shifting job markets. Online platforms make it easier for individuals to continuously learn throughout their lives, whether through formal degree programs, professional development courses, or personal interest pursuits.

Integrating online learning into traditional education systems also has the potential to enhance the quality of education. Educators can supplement in-person instruction with online tools and resources, creating more dynamic and interactive learning experiences. For example, blended learning models combine online learning with face-to-face instruction, allowing students to access resources and materials at their convenience while still benefiting from the guidance and support of a teacher. This hybrid approach can improve student engagement, foster independent learning, and provide a more personalized education experience.

Online learning has revolutionized education delivery, making it more accessible, flexible, affordable, and inclusive. By breaking down geographical, financial, and institutional barriers, online learning has democratized education, offering opportunities for individuals who may not have access to traditional institutions. This has profound implications for the global education systems, as online learning empowers individuals to acquire new skills, pursue their passions, and improve their quality of life. As the world continues to embrace technology and innovation, online learning will play an increasingly vital role in shaping the future of education and ensuring that learning opportunities are available to all, regardless of background or circumstance.

However, these changes also raise concerns about the quality of education, digital divides, and the fear of job displacement in a rapidly changing economy. Education embodies a dual nature, serving as a beacon of hope while generating fear-related inequality and performance pressures. Understanding this duality is essential for fostering an educational environment that nurtures hope, equity, and resilience.

The Duality of Education

Education, in its most fundamental sense, imparts knowledge, skills, and values to individuals. It is regarded as one of the most significant forces for societal transformation and personal empowerment. However, optimism and progress do not define its role in society. Education also holds the potential to invoke fear, anxiety, and exclusion. As both a source of hope and fear, this duality of education can shape the lives of students, educators, and communities. It is essential to recognize, understand, and navigate this dual nature to foster an environment that nurtures hope, equity, and resilience. By doing so, we can ensure that education serves as a vehicle for growth and empowerment rather than a tool of oppression and division.

Education as a Source of Hope

At its best, education is a beacon of hope. It represents the possibility of a better future where individuals have the tools to change their lives, their communities, and the world at large. Education exposes young minds to innovative ideas, cultures, and philosophies. They are encouraged to think critically, ask questions, and challenge assumptions. This intellectual freedom creates opportunities for innovation, personal development, and social progress.

The promise of education is significant for marginalized communities. Education serves as a leveling force, providing individuals with the means to overcome systemic barriers such as poverty, racism, and gender inequality. It opens doors to better economic opportunities and improves healthcare, justice, and social mobility access. For many, education represents the only pathway out of cycles of disadvantage, offering them a chance to rewrite their futures. In this way, education fosters hope for individuals and the broader society, as it empowers citizens to address societal challenges and contribute to collective progress.

Moreover, education nurtures resilience by instilling in students a sense of agency and self-efficacy. Through learning, individuals gain confidence in navigating life's challenges and confronting adversity. This resilience is crucial for personal growth and the development of emotional intelligence. Educated individuals are often better equipped to manage setbacks and failures because they understand that mistakes are part of the learning process and can lead to growth.

Education as a Source of Fear

Despite its potential to inspire hope, education can also be a source of fear. The pressures associated with academic achievement—ranging from standardized testing to the expectations of parents, teachers, and society—can be overwhelming for many students. For some, the fear of failure and the constant need to measure up to others' standards create anxiety that stifles creativity and intellectual curiosity. This fear is particularly pronounced in environments where education is seen solely as a means of securing a job or achieving economic success rather than as a personal growth and exploration tool.

Furthermore, the fear of exclusion and marginalization within the educational system is a harsh reality for many students, especially those from historically underrepresented groups. For students of color, students with disabilities, and LGBTQ+ students, the school environment can be a place of discrimination, bullying, and isolation. Systemic educational inequalities, such as unequal access to quality teachers, resources, and facilities, often compound these fears. In such contexts, education becomes a site of oppression rather than liberation, reinforcing existing power dynamics and perpetuating social divides.

The emphasis on standardized testing and rigid academic benchmarks often exacerbates the fear of failure. These systems fail to account for the diverse ways students learn and grow. The pressure to conform to a narrow set of educational expectations can leave some students feeling inadequate, unworthy, or incapable. This fear can lead to disengagement from the learning process, lower self-esteem, and, in some cases, long-term educational disengagement.

The Need for an Educational Environment that Nurtures Hope, Equity, and Resilience

The duality of education—its potential to inspire both hope and fear—demands a thoughtful, intentional approach to designing educational environments. To cultivate a culture of hope, equity, and resilience, it is crucial to address education's positive and negative aspects and understand how each influences students' experiences and outcomes.

A key component of an equitable educational environment is ensuring all students have access to the resources and opportunities needed to succeed. This includes access to high-quality teachers, safe and inclusive learning spaces, and culturally relevant curricula that reflect students' diverse backgrounds and experiences.

Education must move beyond simply preparing students for the workforce; it must also focus on nurturing their intellectual and emotional growth, encouraging curiosity, critical thinking, and personal development.

Likewise, to reduce the fear of failure, education systems must move away from a one-size-fits-all approach. Stressing personalized learning and recognizing students' different strengths, learning styles, and needs can help reduce anxiety and create a more supportive environment. When students feel that their individuality is recognized and valued, they are more likely to take risks in their learning and embrace challenges rather than avoid them out of fear of failure.

In addition, educators must be trained to recognize and address students' emotional and psychological needs. This includes creating a classroom culture where mistakes are seen as learning opportunities, students are encouraged to be open about their struggles, and where emotional well-being is prioritized alongside academic success. When students feel safe to be vulnerable, they are more likely to develop resilience and a growth mindset, which will serve them well academically and personally.

Creating an environment of hope and equity also requires a commitment to social justice. Education systems must actively dismantle the barriers disproportionately affecting marginalized communities. This means addressing the impact of racism, classism, ableism, and other forms of discrimination that limit opportunities for some students. Schools must be places where all students feel seen, heard, and valued, regardless of their background or identity.

Finally, fostering resilience in students requires emphasizing social-emotional learning (SEL). SEL helps students develop the skills to manage their emotions, set goals, and build healthy relationships. By equipping students with these skills, educators can help them develop the emotional resilience necessary to navigate life's inevitable challenges and setbacks. Moreover, SEL promotes empathy and compassion, assisting students to build stronger connections with one another and creating a more supportive and caring school community.

The duality of education, as both a source of hope and fear, reflects the complexity of human experience. Education can transform lives and societies, offering opportunities for growth, self-empowerment, and social progress. However, it can also be a source of anxiety, exclusion, and oppression, especially for those who are marginalized or whose needs are not adequately met. It is essential to acknowledge this duality and design systems that foster positive emotional and intellectual development for all students to create an educational environment that nurtures hope, equity, and resilience.

By addressing the systemic barriers perpetuating fear and inequality and creating inclusive, supportive, and empowering educational spaces, we can ensure that education remains a tool of liberation and transformation. In doing so, we help individuals realize their full potential and cultivate a society that values hope, resilience, and social justice. Education can be a force that brings people together, empowers communities, and fosters a more equitablc and just world.

Chapter 5

POLITICAL, ECONOMIC, PROMISES AND DISAPPOINTMENTS

The Psychological Impact of Economic Downturns: Fear, Unemployment, and Insecurity

Economic downturns are a natural part of any economy's cycle, characterized by a significant decline in economic activity, such as production, consumption, and employment. They can occur due to various factors, including financial crises, market corrections, political instability, or even pandemics. During these periods of economic hardship, there is a marked increase in unemployment, poverty, and widespread uncertainty. More than just monetary loss, economic downturns lead to significant psychological consequences, including fear, anxiety, and a sense of insecurity. These psychological impacts are particularly pronounced among individuals who experience direct monetary loss, such as those who lose their jobs, face reduced income, or struggle with mounting debt. Understanding the psychological effects of economic uncertainty is crucial for developing strategies to mitigate its negative consequences for individuals and society.

Economic Downturns and Fear

One of the most immediate psychological reactions to an economic downturn is fear (Frasquilho et al., 2016.

The psychological effects of economic downturns, such as stress, anxiety, and depression, are well-documented in research. For example, a systematic review published in *BMC Public Health* highlights the association between economic recessions and mental health challenges, including increased rates of common mental disorders and substance-related issues. Another study in the *Epidemiology and Psychiatric Sciences* journal discusses how economic crises can lead to higher demand for mental health care and increased use of prescription drugs. These findings underscore the significant impact of financial instability on mental well-being. Fear arises from the uncertainty about the future, particularly regarding personal finances, job security, and the ability to maintain a certain standard of living. During economic hardship, people experience a heightened sense of vulnerability, knowing that they may lose their jobs or face a drastic reduction in income. This fear of job loss and financial instability is a significant source of stress, affecting not only individuals but also families, communities, and even entire nations.

Fear is particularly acute when the economic downturn is unexpected or severe, as was the case during the 2008 global financial crisis and the 2020 COVID-19 pandemic. Both events created a perfect storm of job losses, business closures, and widespread uncertainty. For individuals, fear manifests as a constant worry about the future. Will they be able to pay their bills? Will they be able to support their families? For many, these fears become all-consuming, leading to sleep disturbances, difficulty concentrating, and even physical symptoms such as headaches or stomach issues.

This constant state of fear can also lead to the development of chronic stress, which has been linked to a wide range of mental and physical health problems. According to the American Psychological Association (APA), stress from economic hardship can contribute to anxiety disorders, depression, substance abuse, and cardiovascular issues. Fear is often compounded by the social stigma associated with unemployment, as individuals may feel embarrassed or ashamed of their financial situation, further intensifying their psychological distress.

Unemployment and Its Psychological Effects

Unemployment is one of the most direct consequences of an economic downturn and has profound psychological effects on individuals. Being unemployed often leads to feelings of worthlessness, isolation, and a loss of identity. For many people, their job is not just a source of income but also a key part of their self-concept and social identity. Losing a job can, therefore, feel like a personal failure, leading to low self-esteem. This can be especially difficult for individuals who have been in their careers for an extended period or have never experienced unemployment.

The psychological effects of unemployment are well-documented. Studies have shown that unemployed people are more likely to experience depression, anxiety, and a variety of other mental health issues. For instance, a study published in the "American Journal of Epidemiology" found that unemployment is strongly associated with increased rates of depression and anxiety, with the effects being more pronounced among those who have been unemployed for more extended periods.

The impact of unemployment is not limited to the individual level. The psychological strain of unemployment often affects families, especially if it results in financial hardship. Unemployed parents may experience a sense of guilt for not being able to provide for their children, leading to further anxiety and stress. Children in such families may also experience emotional distress due to changes in family dynamics, such as increased tension or the loss of a sense of security.

Long-term unemployment can lead to social isolation (Mannl,2012). As individuals struggle to find new work, they may withdraw from social activities and avoid interactions with friends and family, feeling ashamed of their situation. This isolation can exacerbate feelings of loneliness and depression, making it more difficult for individuals to recover emotionally and psychologically.

Also, in times of economic downturn, insecurity becomes a pervasive feeling. Economic instability threatens a person's fundamental pillars of stability, including housing, health care, and personal safety. For many individuals, the loss of a steady income creates a sense of insecurity about their ability to maintain a basic standard of living. This can lead to various emotional responses, from anger and frustration to fear and helplessness.

Oftentimes, insecure individuals may worry about the possibility of losing their homes, being unable to afford health care, or facing a future of persistent financial hardship. Rising living costs, including inflation and the increasing costs of food, utilities, and other necessities, often exacerbate this instability. As a result, people begin to feel that they are at the mercy of external forces beyond their control, leading to a loss of agency and empowerment. Insecurity can also trigger feelings of hopelessness, which can be particularly dangerous when individuals begin to perceive their economic situation as inescapable. Insecurity is not just an individual experience but has a broader societal impact. When large segments of the population are insecure, it can lead to social unrest, political instability, and increased crime rates. People who feel economically insecure are more likely to take drastic actions, whether turning to criminal activity to survive or protesting the system they believe is failing them. In some cases, feelings of insecurity can lead to the breakdown of social cohesion, as people become more focused on their survival rather than on collective well-being.

Coping Mechanisms and Psychological Resilience

While the psychological effects of economic downturns can be profound, many individuals and communities exhibit remarkable resilience. Coping mechanisms play a crucial role in how people respond to economic hardship. Social support, for example, can be a vital resource for those facing unemployment or financial insecurity. A strong network of family, friends, or community organizations can help individuals weather the psychological storm of economic downturns.

Additionally, psychological interventions such as counseling and therapy can provide individuals with tools to cope with anxiety, depression, and stress. For example, Cognitive-Behavioral Therapy (CBT) can help individuals reframe negative thoughts, reduce stress, and develop healthier coping strategies. In some cases, economic downturns can lead to personal growth, as individuals may find new sources of meaning or reevaluate their values and priorities.

Governments and organizations can also help alleviate the psychological burden of economic uncertainty. Social safety nets, such as unemployment benefits, healthcare coverage, and food assistance programs, can help mitigate the financial strain caused by job loss and economic insecurity. By providing a basic level of security, these programs allow individuals to focus on rebuilding their lives rather than constantly worrying about their immediate survival.

Economic downturns are not just a financial or statistical issue; they are deeply psychological events that can profoundly alter the lives of individuals and communities. The fear, unemployment, and insecurity that arise during these periods can have long-lasting effects on mental health, leading to increased rates of anxiety, depression, and social isolation. Understanding the psychological impacts of economic uncertainty is crucial for developing effective interventions that can alleviate suffering and support individuals as they navigate challenging financial times.

Societies can mitigate the psychological toll of economic downturns by fostering resilience, strengthening social support systems, and ensuring access to mental health care. While the future will inevitably present more challenges, understanding and addressing the psychological impacts of economic uncertainty is vital to ensuring that individuals can recover and thrive in the face of adversity.

How might we bridge the divide between hope and fear amid economic challenges? Some ways are through entrepreneurship and innovation. Financial challenges, whether driven by global recessions, technological disruption, or the unpredictability of market forces, can profoundly impact individuals, businesses, and societies. These challenges often bring about feelings of uncertainty and despair. However, amidst such adversity, innovation and entrepreneurship emerge as powerful forces capable of revitalizing economies, empowering communities, and offering hope for a brighter future. Let us explore how innovation and entrepreneurship can foster hope in the face of economic challenges, with a particular focus on the potential of these forces to drive economic recovery, create opportunities, and inspire social change.

1. Understanding Economic Challenges

Before examining the role of innovation and entrepreneurship, it is essential to understand the nature of economic challenges. These challenges can take many forms, including high unemployment, stagnating wages, rising inequality, and economic recessions.

Economic downturns, such as the global fiscal crisis of 2008 or the COVID-19 pandemic, have demonstrated how quickly economic stability can be undermined, leading to widespread hardship.

During such times, traditional economic systems and policies may struggle to adapt, often disillusioning individuals and communities. As a result, the pursuit of financial stability and progress becomes more complex. However, during these challenging times, innovation and entrepreneurship can provide transformative solutions to help rebuild and reposition economies for the future.

2. The Role of Innovation in Economic Recovery

Innovation refers to the process of developing innovative ideas, methods, or products that create value and solve problems in novel ways. In the context of economic challenges, innovation plays a critical role in the recovery process by driving productivity, improving efficiency, and opening new avenues for growth.

One key area where innovation has been instrumental is in the realm of technology. Technological advancements, such as the development of the internet, artificial intelligence, and renewable energy solutions, have revolutionized industries, creating new markets and opportunities for employment. In times of economic downturn, these innovations can function as catalysts for recovery by creating entirely new industries, reducing reliance on outdated or declining sectors, and offering solutions to longstanding social problems. For example, the rapid expansion of the technology sector during the early 21st century has provided opportunities for individuals with specialized skills, leading to a more diversified workforce.

Moreover, innovation often creates more efficient processes and products, allowing businesses to reduce costs and increase profitability. This, in turn, can lead to job creation and economic growth. For instance, during the 2008 fiscal crisis, the rise of online services and digital platforms provided new economic opportunities, particularly for entrepreneurs, freelancers, and small businesses.

3. Entrepreneurship as a Pathway to Economic Empowerment

Entrepreneurship, which is starting and managing new businesses, is intrinsically tied to innovation. Entrepreneurs are often the ones who identify market gaps, take risks, and bring the latest ideas to life. In times of economic hardship, entrepreneurship can serve as a pathway to economic empowerment by offering individuals the chance to create their own jobs, contribute to local economies, and inspire others to do the same. One of the key ways that entrepreneurship fosters hope amid economic challenges is through job creation. In many economies, small and medium-sized enterprises (SMEs) are the backbone of job creation. According to the World Bank, SMEs generate many new jobs globally. In times of high unemployment, entrepreneurs can help reduce the economic strain by hiring individuals, providing them with income, and offering career development opportunities. By creating their own businesses, entrepreneurs help to alleviate the burden on public welfare systems and contribute to a more resilient economy. Entrepreneurship also empowers individuals by providing them with autonomy to shape their futures. Rather than relying solely on traditional employment opportunities, which may be scarce during economic crises, individuals can take matters into their own hands by starting businesses that reflect their passions and expertise. This sense of control and self-determination can offer a powerful sense of hope, particularly for those who may have lost their jobs or faced limited career prospects in the traditional labor market.

Furthermore, entrepreneurship fosters a spirit of resilience and adaptability. Entrepreneurs must be resourceful and able to pivot when faced with challenges. This mindset is crucial during economic difficulty, as businesses must find ways to innovate and adapt to changing circumstances. The ability to bounce back from failure, learn from mistakes, and stay motivated in the face of adversity is a characteristic shared by many successful entrepreneurs. This resilience can inspire hope in others, demonstrating that setbacks are not permanent and that opportunities can arise even in the most challenging of times.

4. Social Impact and Community Building

Beyond economic recovery, innovation and entrepreneurship can also have profound social implications, helping to build stronger, more inclusive communities. Many entrepreneurs focus on solving social issues and creating businesses that address the needs of marginalized or underserved populations. These social entrepreneurs not only contribute to economic growth but also foster a sense of purpose and unity within society.

For example, businesses that focus on sustainability, education, healthcare, and social equity can make a tangible difference in the lives of individuals struggling with economic inequality's effects. Social innovation can lead to developing products or services that improve the quality of life, increase access to essential services, and promote social inclusion. In addition, entrepreneurship can strengthen community bonds by fostering local economic development. Small businesses that are locally owned and operated often have a deeper connection to their communities, and their success is closely tied to the well-being of the people they serve. As entrepreneurs reinvest in their local communities, they help to create a cycle of economic vitality and social cohesion.

5. Inspiration and Hope for the Future

At a psychological level, innovation and entrepreneurship can foster a sense of hope by inspiring individuals to believe that change is possible. In times of economic distress, it is easy to feel overwhelmed by the challenges that lie ahead. However, when people see entrepreneurs and innovators creating new businesses, launching innovative technologies, and developing solutions to pressing problems, it reinforces the idea that progress is attainable.

The stories of successful entrepreneurs who have risen from humble beginnings to achieve remarkable success can be powerful motivators. They show that with determination, creativity, and hard work, individuals can overcome obstacles and make a significant impact. These success stories not only inspire others to pursue their entrepreneurial dreams but also remind society that hope is not just about surviving economic challenges but about finding opportunities to thrive and contribute to the greater good.

Moreover, as innovation and entrepreneurship continue to drive societal progress, they instill a sense of optimism about the future. By solving problems related to climate change, poverty, healthcare, and education, entrepreneurs are shaping a more sustainable, equitable, and connected world. This forward-thinking mindset can help shift the focus from short-term economic struggles to long-term opportunities for growth and development.

Innovation and entrepreneurship play a crucial role in fostering hope amid economic challenges. Through their ability to drive recovery, create opportunities, empower individuals, and address social issues, these forces provide a pathway for societies to navigate tough times. While economic challenges are an inevitable part of human experience, the resilience, creativity, and determination exhibited by innovators and entrepreneurs demonstrate that adversity can be overcome and that a brighter future is within reach. By embracing innovation and entrepreneurship, we can build stronger, more resilient economies and communities where hope flourishes despite hardship.

Globalization and Economic Disparities

Globalization has created both hope and fear in the economic landscape. While it offers opportunities for growth and collaboration, it raises concerns about job displacement and widening economic disparities. The fear of losing jobs to globalization can lead to protectionist sentiments and social tensions. Analyzing the complexities of globalization, including its impact on local economies and labor markets, can provide valuable insights into the dual nature of hope and fear in a globalized world.

The relationship between economic factors, hope, and fear is intricate and multifaceted. While financial stability can foster hope, downturns and disparities can generate significant fear. Understanding these dynamics is crucial for developing strategies to enhance resilience and promote a hopeful financial future.

Chapter 6

RELIGIOUS BELIEFS AND DOUBTS

The role of religion in providing hope and instilling fear

Religious beliefs play a profound role in shaping human responses to uncertainty, offering both solace and caution. They provide hope by offering believers a sense of purpose, community support, spiritual practices, and the promise of redemption. Many religions emphasize the importance of personal commitment and renewal, forgiveness, and the possibility of a better future, which can inspire optimism. At the same time, spiritual beliefs contribute to the emergence of a sense of fear in the admonition for moral accountability, the use of apocalyptic narratives to underscore religious beliefs, and the perceived requirement of social conformity. This duality—hope, and fear—makes religion a powerful force in navigating uncertainty. It can inspire resilience and unity but also provoke anxiety or division, depending on how beliefs are interpreted and applied.

Throughout history, religion has often been a pillar of hope during tumultuous times. For example, religious narratives have offered solace and explanations for suffering during war and plagues. Various religious figures, such as Martin Luther King Jr., have inspired hope through their messages of justice and equality. Sacred texts provide guidance and comfort, reminding followers of a higher purpose and the potential for redemption. Analyzing specific historical events, such as the Black Death in Europe and the role of the Church in providing community support, illustrates the multifaceted ways religion can nurture hope.

Fear of the unknown:

Human history is rich with ideas, beliefs, and philosophies that have shaped societies and individuals alike. Among these, a few concepts have been as pervasive and influential as *existential dread*, the fear of divine judgment, and the role of dogma (Kierkegaard,1846). These elements have coexisted and interacted in various religious, philosophical, and cultural systems, often in ways that have profoundly influenced how people understand their existence, the cosmos, and their fate. This essay explores the historical development of these intertwined concepts, examining how existential dread has been expressed through religious and philosophical thought, how the fear of divine judgment has shaped human behavior, and the significant role dogma has played in providing comfort and constraint.

At the heart of existential dread lies the human confrontation with the meaninglessness of existence. This dread can be understood as the fear of facing a world that is indifferent or hostile to human life. Existential philosophers, such as Søren Kierkegaard, Friedrich Nietzsche, and Jean-Paul Sartre, have grappled with these themes, often finding that individuals are thrown into a world where meaning must be created, not found.

Kierkegaard, often regarded as the father of existentialism, emphasized the individual's struggle with despair and anxiety in the face of existence. For him, existential dread stemmed from the realization of human finitude and the uncertainty of one's relationship with God. Kierkegaard's (1846) concept of "the leap of faith" posited that only by embracing religious faith could individuals overcome this dread, suggesting that belief in God offers a way to confront the abyss of meaninglessness.

Nietzsche, on the other hand, famously proclaimed that "God is dead," marking the decline of traditional religious beliefs in the modern world. This declaration underscored the existential crisis of the modern age, in which the old certainties of divine meaning and purpose no longer provided a framework for understanding life. Nietzsche believed that individuals must face the abyss of a godless universe and create their own values through the "will to power." His philosophy of the Übermensch (superman) called for self-overcoming and the creation of meaning in a world that offered none intrinsically.

In the 20th century, existentialism continued to grapple with these themes. Sartre's notion of "bad faith" explored how individuals deceive themselves in the face of freedom and responsibility. According to Sartre, humans are condemned to be free, meaning that they bear the weight of choosing their existence and meaning without relying on preordained structures. Existential dread, therefore, becomes a central part of the human condition as individuals realize the profound responsibility they bear in creating meaning in an indifferent universe.

Theological Perspectives on Human Fate

The fear of divine judgment has been a central theme in many religious traditions. From the ancient gods of Greek and Roman mythology to the monotheistic traditions of Christianity, Judaism, and Islam, the fear of judgment by a higher power has shaped human behavior and moral codes. This fear is often intertwined with existential dread as individuals confront the possibility of divine punishment or reward for their actions in life.

In Christianity, the concept of divine judgment is most clearly articulated in the doctrine of the Last Judgment, in which God will judge all souls based on their actions during their lifetime. This judgment determines their eternal life in heaven or eternal damnation in hell. The fear of divine judgment, particularly the fear of hell, has historically played a significant role in shaping Christian morality. The concept of original sin, the fall of man, and the necessity of redemption through faith in Christ have all contributed to a theological framework in which individuals are constantly aware of their moral shortcomings and the need for divine grace.

Similarly, in Islam, the fear of divine judgment is integral to the faith. The Qur'an repeatedly emphasizes the Day of Judgment, where individuals will be held accountable for their deeds. In Islam, salvation is achieved through a combination of faith in God and charitable deeds, with the fate of the individual being determined by God's will. The fear of divine judgment is a powerful motivator for ethical behavior and religious observance.

In contrast to these religious views, some philosophical traditions, such as those of ancient Greece, did not emphasize divine judgment similarly. Greek philosophy, particularly that of the Stoics, focused more on living in accordance with nature and reason rather than the fear of divine retribution. Nevertheless, even in these traditions, there was an acknowledgment of fate and the role of divine forces in shaping human destiny.

The Role of Dogma: Providing Certainty and Structure

Dogma, as a set of established beliefs or doctrines, has played a crucial role in shaping human understanding of existential dread and the fear of divine judgment. Dogmatic systems, whether religious or philosophical, provide a framework within which individuals can make sense of their lives and the world around them. In this sense, dogma offers comfort, certainty, and stability in the face of existential anxiety.

In religious contexts, dogma often serves as a means of providing answers to fundamental questions about the nature of existence, the divine, and the afterlife. Christianity, for example, has a well-established set of dogmas regarding the nature of God, the Trinity, the resurrection, and the salvation of souls. These dogmas offer a clear and structured understanding of the world and the individual's place within it. They provide a sense of security by answering questions that might otherwise lead to existential dread.

Similarly, Islam has a set of dogmatic beliefs that define the nature of God (Allah), the role of the Prophet Muhammad, and the path to salvation. These dogmas provide a framework within which Muslims can navigate the uncertainties of life and death, offering a sense of purpose and direction.

Dogma, however, is not without its criticisms. The rise of secularism and scientific rationalism in the modern era has led many to question the validity of religious dogma. Thinkers such as Karl Marx, Sigmund Freud, and Friedrich Nietzsche criticized dogmatic belief systems, viewing them as tools of social control or psychological repression. Marx argued that religion was the "opium of the people," a means of pacifying the masses and preventing them from confronting the injustices of their material conditions. Freud saw religion as a projection of human desires for safety and security in the face of an uncertain world. Nietzsche, as mentioned earlier, rejected dogma as a constraint on individual freedom and self-creation.

Despite these critiques, dogma continues to play a significant role in shaping religious and philosophical systems. For many individuals, adherence to dogmatic beliefs provides a sense of meaning and purpose that helps mitigate existential dread. Dogma offers a clear narrative for understanding the human condition, the nature of the divine, and the consequences of human action.

Existential dread, the fear of divine judgment, and the role of dogma have been central themes in the history of human thought. These concepts have shaped the way people approach life, death, morality, and the meaning of existence. While existential philosophers have highlighted the anxiety and responsibility that comes with human freedom, religious traditions have provided a framework for understanding the divine and the consequences of human actions. The tension between existential dread and the fear of divine judgment resonates today as individuals seek meaning and purpose in an often chaotic and uncertain world. In a modern context, secularism, scientific rationalism, and the questioning of religious dogma have altered how people grapple with these issues. Yet, for many, the fear of divine judgment and the comfort of dogmatic beliefs remain central to their understanding of existence.

As we move further into the 21st century, the interplay between faith, reason, and fear will continue to shape our collective consciousness. Whether through the lens of existentialism, theology, or the critique of dogma, the search for meaning in the face of existential dread and the fear of divine judgment remains one of the most profound and enduring questions in the history of human thought.

Hope through Faith

Faith-based practices such as prayer, meditation, and community gatherings can foster hope and resilience in the face of uncertainty. These practices help individuals cope with anxiety, providing a sense of belonging and support. Religious communities often mobilize to help those in need, reinforcing the idea that collective faith can overcome adversity. Additionally, religious teachings frequently emphasize themes of hope, renewal, and the possibility of change, encouraging followers to remain optimistic even during challenging times. Exploring case studies of religious groups who have successfully navigated crises can illustrate the power of faith in fostering hope.

Consider the impact of modern issues (e.g., secularism, interfaith conflicts) on religious hope and fear and how these challenges impact our lives. In contemporary society, religious institutions face challenges such as secularism, interfaith conflicts, and declining membership. These factors can create uncertainty within religious communities, leading to fear about the relevance of faith in an increasingly secular world. However, many religious organizations are adapting by embracing social justice movements, addressing contemporary issues like climate change, and fostering interfaith dialogue. This adaptability can generate hope, as it demonstrates a commitment to evolving and addressing the needs of society. Discussing examples of successful religious initiatives that engage with modern challenges can highlight this dynamic.

Religious organizations have been integral to societies throughout history, providing moral guidance, communal cohesion, and spiritual support. However, with the rapid pace of social, cultural, and technological changes in the modern world, religious groups have confronted an array of uncertainties. These uncertainties, ranging from societal shifts and scientific advancements to political instability and global crises, pose challenges to the traditional structures and teachings of religious organizations. In response, many religious groups have undergone significant transformations, either by adapting existing practices or introducing new paradigms to meet the needs of contemporary adherents.

This chapter explores how religious organizations have navigated the complexities of contemporary uncertainties. By examining case studies from various religious traditions, including Christianity, Islam, Judaism, and Buddhism, we can gain insight into the methods employed by religious leaders to ensure their relevance and continuity in an ever-changing world. The essay will also delve into the broader implications of these adaptations for religious communities and their place in society.

The Role of Religious Organizations in Times of Uncertainty

In times of uncertainty, religious organizations often serve as pillars of support and resilience for individuals and communities. They provide a sense of stability through shared beliefs, rituals, and moral guidance, helping people navigate challenges with hope and purpose. Beyond spiritual solace, these organizations frequently offer practical assistance, such as food, shelter, and counseling, fostering a sense of community and belonging. By addressing both emotional and material needs, religious institutions play a vital role in helping individuals find meaning and strength during periods of crisis or upheaval (Kim,2020).

These references provide a foundation for understanding the multifaceted roles of religious organizations in supporting individuals and communities during challenging times.

To understand how religious organizations adapt, examining the nature of contemporary uncertainties is important. These uncertainties can be categorized into several broad areas:

1. **Social and Cultural Changes**: Globalization, migration, and the rise of secularism have led to significant shifts in societal norms and values. Traditional religious teachings often clash with modern concepts such as gender equality, LGBTQ rights, and personal freedom. This creates tension within religious communities as they struggle to reconcile traditional beliefs with contemporary ethical standards.

2. **Scientific and Technological Advancements**: The rapid development of science and technology has raised questions about the existence of a divine creator, the nature of the soul, and the ethical implications of modern technologies, such as genetic engineering and artificial intelligence. These advancements challenge religious doctrines and force religious organizations to engage in dialogue with the scientific community.

3. **Political Instability and Conflict**: In many parts of the world, religious organizations find themselves at the center of political struggles, whether they are advocating peace, justice, or human rights. Conflicts between different

religious groups and between religion and state create a need for religious leaders to engage in political and social activism while remaining true to their core beliefs.

4. **Environmental and Global Crises**: The ongoing climate crisis, along with other global challenges such as pandemics, economic inequality, and migration, has called on religious organizations to take a more active role in addressing these issues. Many religious groups have recognized the importance of environmental stewardship, social justice, and humanitarian efforts, considering the growing number of global crises.

Adapting to Social and Cultural Changes

Religious organizations have responded to social and cultural changes in many ways. One of the most significant challenges religious institutions face is the changing role of women in society. Traditional religious views on gender roles often conflict with modern feminist ideals, leading to debates within religious communities about the role of women in leadership positions and their participation in religious rituals.

For instance, in many branches of Christianity, especially the Roman Catholic Church, the question of women's ordination remains a contentious issue. While some denominations have embraced female clergy, others maintain that the priesthood is reserved for men based on traditional interpretations of Scripture. In response to the evolving role of women in society, many religious organizations have had to reconsider their stance on gender equality. Some Protestant denominations, for example, have made significant strides by ordaining women and allowing them to serve as pastors, bishops, and other leadership roles.

Similarly, Islamic organizations have been forced to address issues related to women's rights, including education, employment, and participation in public life. Although traditional interpretations of the Quran emphasize patriarchal values, there has been a growing movement within the Muslim world to reinterpret Islamic teachings in ways that promote gender equality. Women's leadership in religious spaces has been increasingly acknowledged, with some

Muslim women assuming positions of authority in both religious and political spheres.

LGBTQ rights present another area where religious organizations have had to adapt. While many religious traditions have historically condemned same-sex relationships, a growing number of religious groups have re-examined their positions, considering evolving social attitudes. Some Christian denominations, such as the United Church of Christ and the Episcopal Church, have embraced LGBTQ rights and have allowed for same-sex marriages and the ordination of LGBTQ clergy. Other religious organizations, however, continue to uphold traditional views on marriage and sexuality, creating divisions within communities and broader societal debates on religious tolerance.

Engaging with Science and Technology

The relationship between religion and science has long been a source of tension, but contemporary religious organizations have increasingly sought ways to engage with scientific advancements. One of the most pressing issues in this regard is the theory of evolution, which challenges traditional creationist views. While some religious groups continue to reject evolution, others have adopted more flexible interpretations of sacred texts, suggesting that religious teachings and scientific theories are not necessarily in conflict.

For example, the Catholic Church, which historically opposed the theory of evolution, has since adopted a more conciliatory stance. In 1996, Pope John Paul II acknowledged that evolution is "more than a hypothesis" and encouraged dialogue between science and religion. This shift reflects a broader trend within religious organizations to embrace scientific findings without compromising core religious beliefs.

In addition to issues related to evolution, technological advancements in fields like artificial intelligence and genetic engineering have raised ethical dilemmas for religious communities. For instance, the rise of gene editing technologies such as CRISPR has prompted religious leaders to consider questions about the sanctity of life, the morality of altering the human genome, and the potential for "designer babies." Religious organizations, particularly in the Abrahamic faiths, have expressed concerns about the ethical implications of these technologies and have called for the establishment of moral guidelines for their use.

Buddhism, with its emphasis on the interconnection of all beings, has also engaged with contemporary technologies. Buddhist leaders have expressed interest in the ethical implications of artificial intelligence, arguing that technology should be used to promote compassion and reduce suffering rather than exacerbate societal inequalities.

Responding to Political Instability and Conflict

Religious organizations often play a key role in addressing political instability and conflict. Whether it is through interfaith dialogue, humanitarian work, or political activism, religious leaders have long been involved in efforts to promote peace and justice. In regions of the world affected by political turmoil, such as the Middle East and Sub-Saharan Africa, religious organizations often serve as mediators, helping to broker peace agreements and providing support to those affected by violence and displacement.

Religious organizations have sometimes been at the forefront of political movements. The Catholic Church, for example, played a significant role in the Solidarity movement in Poland, which helped bring about the fall of communism in Eastern Europe. In Latin America, liberation theology emerged in the 20th century as a response to social and political injustices, with religious leaders advocating for the rights of the poor and marginalized.

However, religious organizations are not always neutral actors in political conflicts. In many parts of the world, religious identity is deeply intertwined with political power, and religious groups can become embroiled in political struggles. This is evident in the ongoing conflicts between Sunni and Shia Muslims in the Middle East, where religious organizations are often aligned with political factions. Similarly, the rise of religious nationalism in countries like India, Israel, and the United States has led to tensions between religious groups and political authorities.

Addressing Environmental and Global Crises

As global crises such as climate change, pandemics, and economic inequality continue to intensify, religious organizations have become increasingly active in addressing these issues. Many religious traditions have long advocated for environmental stewardship, with teachings that emphasize the importance of caring for the Earth. Pope Francis' 2015 encyclical, *Laudato Si'* called for a global response to climate change, urging people of all faiths to take responsibility for the environment and work towards sustainable development.

Similarly, Buddhist teachings on compassion and interconnectedness have led to an emphasis on environmental sustainability. The Dalai Lama has repeatedly spoken out on the need to protect the planet, and many Buddhist organizations have launched initiatives aimed at reducing environmental harm and promoting ecological balance.

Religious organizations also play a vital role in addressing social and economic inequalities exacerbated by global crises. Catholic charities, Islamic relief organizations, and Jewish humanitarian groups have been active in providing aid to refugees, combatting poverty, and advocating for social justice. These efforts often stem from religious teachings that emphasize compassion, charity, and the inherent dignity of all people.

Religious organizations face numerous challenges as they navigate the complexities of contemporary uncertainties. From social and cultural shifts to scientific advancements, political instability, and global crises, religious groups must adapt to remain relevant and effective in their mission. While some religious organizations have resisted change, others have embraced innovative ideas and practices to address the needs of their communities. In doing so, they not only preserve their traditions but also contribute to the broader conversation about the role of religion in a rapidly changing world.

As the world continues to evolve, religious organizations will continue to face new challenges and opportunities for adaptation. Their ability to engage with contemporary uncertainties while staying true to their core values will determine their continued relevance and impact in the 21st century.

The dual role of religion as a source of hope and fear is complex and multifaceted. While faith can provide comfort and a sense of purpose, it can also lead to anxiety and existential dread. Understanding this duality is crucial for recognizing how religion shapes human experience during uncertain times.

Chapter 7

THE INTERCONNECTEDNESS AMONG ECONOMICS, EDUCATION AND RELIGION

Religion, economics, and education are deeply interconnected, each influencing the others in significant ways. For instance, religious institutions often provide educational opportunities and economic support to communities. Examining examples of faith-based organizations that offer scholarships, vocational training, or economic development initiatives can illustrate the collaborative efforts across these domains. Understanding these interconnections can highlight society's holistic nature of hope and fear.

In the 21st century, two of the most significant and interconnected challenges the world faces are political instability and climate change. Both issues pose not only direct threats to national security and global well-being but also often exacerbate one another, creating a vicious cycle of instability and environmental destruction. Political instability and climate change are multi-dimensional issues that involve complex relationships between governance, economic policies, environmental sustainability, and social systems. This chapter explores the familiar challenges of political instability and climate change, examining how these two domains interact and influence each other.

Political instability is characterized by frequent changes in leadership, social unrest, government collapse, or ineffective governance, often resulting in widespread insecurity, violence, or economic deterioration. It arises from various causes, including poor governance, corruption, human rights abuses, internal conflict, and inadequate responses to socio-economic inequalities. The challenges of political instability are most evident in fragile or failing states, where government institutions are either weak or absent. In these contexts, political instability often leads to the collapse of public services, the undermining of economic structures, and the disruption of social order. Countries like Syria, Afghanistan, and Venezuela are examples of nations that have experienced significant political instability, leading to human suffering, displacement, and loss of life. The lack of stability makes it difficult for governments to address critical issues like education, healthcare, and infrastructure while also making it harder for international bodies to intervene effectively.

The consequences of political instability are far-reaching. It disrupts economic growth, creates environments ripe for corruption and violence, and reduces a country's ability to respond to crises—be they health, economic, or environmental. Furthermore, political instability can lead to the erosion of civil liberties and rights, as governments, in their attempt to maintain control, may resort to authoritarian measures or violent repression.

Climate Change: Challenges and Global Implications

Climate change is the long-term alteration of global weather patterns, primarily driven by human activities such as burning fossil fuels, deforestation, and industrialization. The consequences of climate change include rising global temperatures, extreme weather events, loss of biodiversity, sea-level rise, and shifts in ecosystems, all of which significantly impact both natural environments and human populations.

One of the primary challenges of climate change is its far-reaching nature. Climate change affects every aspect of human life, from agriculture and food security to migration and health. For example, extreme heat waves and droughts can decimate crops, leading to food shortages and economic loss. The increasing frequency and severity of hurricanes, floods, and wildfires have devastating impacts on communities, particularly in vulnerable regions.

One of the most pressing challenges of climate change is its disproportionate impact on poorer and less developed countries. These nations, which contribute the least to global greenhouse gas emissions, are often the ones that suffer the most from the effects of climate change. Sub-Saharan Africa, Southeast Asia, and small island states are particularly vulnerable, experiencing extreme weather, rising sea levels, and dwindling resources. Climate-induced migration has become a pressing issue, with many people being displaced from their homes due to flooding, storms, or droughts.

The Intersection of Political Instability and Climate Change

While political instability and climate change are significant challenges on their own, their intersection can magnify their impact, creating new and complex vulnerabilities. For instance, countries experiencing political instability may find it challenging to address the impacts of climate change effectively. Governance systems that are weak or corrupt are often incapable of implementing adaptive policies or mitigating climate-related risks. In these situations, climate change can further destabilize an already fragile political environment, leading to increased conflicts over resources like water, land, and food.

One of the most visible intersections of climate change and political instability is in the context of conflict over natural resources. As climate change depletes vital resources, competition for access to water, arable land, and energy sources becomes more intense. This has already been witnessed in regions such as the Sahel in Africa, where droughts and desertification have led to violent clashes between different ethnic and social groups over land and water. In these cases, political instability can be exacerbated by climate change, as governments fail to manage the consequences of environmental stress and as social groups vie for limited resources.

In other cases, climate change can trigger mass migration, further destabilizing fragile political environments. People displaced by climate-related disasters often seek refuge in neighboring countries or cities, creating pressure on host governments and infrastructures. The influx of climate migrants can fuel xenophobia, social unrest, and conflict, especially when the receiving nation is politically unstable or lacks the resources to support the newcomers. In regions where political structures are already weak, such as the Middle East or parts of Africa, the combination of climate-induced migration and instability can create a dangerous feedback loop, leading to further conflict and suffering.

Global Responses to Political Instability and Climate Change

Addressing the challenges of political instability and climate change requires coordinated global action, as both issues transcend national borders. One of the primary challenges in tackling these two issues together is the need for a multi-level governance approach that includes local, national, and international responses.

International bodies such as the United Nations have attempted to address both political instability and climate change through various initiatives. For example, the United Nations Framework Convention on Climate Change (UNFCCC) and the Paris Agreement aim to limit global warming and reduce carbon emissions. These initiatives are vital for mitigating the long-term impacts of climate change. However, their success depends on the cooperation of governments worldwide, many of which are politically unstable or resistant to change.

Likewise, international organizations and governments have also focused on addressing political instability through peacekeeping missions, development assistance, and diplomatic efforts. The European Union, the United States, and other regional powers often play key roles in promoting stability in politically unstable countries by offering aid, supporting democratic reforms, and fostering diplomatic dialogue. However, political instability in many regions complicates the delivery of aid and support, making it difficult to achieve long-term stability and sustainable development.

In conclusion, the challenges of political instability and climate change are both complex and interconnected, with far-reaching consequences for individuals, communities, and nations. While political instability often exacerbates the effects of climate change by hindering effective governance and exacerbating resource scarcity, climate change, in turn, contributes to political instability by creating environmental stress, economic hardship, and social unrest. Addressing these challenges requires a comprehensive approach that involves coordinated action at the local, national, and global levels. Only through international cooperation, effective governance, and a commitment to sustainable development can the world hope to overcome the dual challenges of political instability and climate change and work toward a more stable, resilient, and equitable future for all.

Many challenges faced by religion, education, and economics during times of uncertainty are shared. Issues such as political instability, social inequality, and climate change affect all three domains, creating a collective sense of fear and urgency. Discussing how these challenges intersect and influence public perceptions can illuminate the complexities of hope and fear in contemporary society.

Community Responses

Communities often rise to meet uncertainties and challenges by mobilizing across religious, educational, and economic dimensions, creating a unified front for resilience and progress. Religious institutions provide moral guidance and emotional support, fostering hope and solidarity. Educational organizations equip individuals with knowledge and critical thinking skills, empowering them to address complex issues. Economic initiatives, such as local businesses and financial programs, drive resource allocation and sustainable development. Together, these dimensions form a dynamic network that enables communities to adapt, innovate, and thrive in the face of adversity (Clauss-Ehlers, Chiriboga, Hunter, Roysircar, & Tummala-Narra, 2019; American Psychological Association, 2017; American Psychological Association Presidential Task Force on Educational Disparities, 2012).

These references provide a foundation for understanding how communities collaborate across various dimensions to address challenges.

The Future of Hope and Fear: Economics, Education, and Religion

In human civilization's vast and unpredictable landscape, hope and fear have long served as the twin engines driving progress and restraint. As humanity moves deeper into the 21st century and beyond, the interconnected domains of economics, education, and religion will continue to evolve in ways that challenge and redefine these fundamental emotions. This speculative narrative seeks to explore the trajectories of hope and fear in these spheres, tracing the possible futures that lie ahead.

Economics: The Dawn of Decentralized Hope and Fear

The global economic system has undergone a seismic transformation with the rise of decentralized finance (DeFi), artificial intelligence, and automation. By the mid-21st century, traditional financial institutions may have lost their dominance to decentralized networks that operate autonomously, managed by smart contracts and blockchain technology. For many, this heralds a new age of financial inclusivity, where economic hope is fueled by greater access to wealth-generation tools and reduced dependence on centralized banks.

However, fear also lurks in this new economic paradigm. The displacement of traditional jobs by AI-driven automation raises concerns about mass unemployment, social stratification, and a growing wealth gap. Even as universal basic income (UBI) and digital currencies attempt to stabilize economies, there is an underlying anxiety about the erosion of traditional economic structures and the uncertainty of employment in a world where human labor is increasingly obsolete.

Hope persists, however, in the form of cooperative economic models, where resource-sharing, sustainability-driven industries, and digital barter systems create a more equitable global economy. Whether these models will successfully replace the old guard or merely coexist with oligarchic wealth remains an open question, balanced precariously on the edge of hope and fear.

Education: The Evolution of Knowledge and the Uncertainty of Purpose

The education sector, long considered the bedrock of societal advancement, is poised for radical change. By 2050, personalized AI tutors, virtual reality (VR) classrooms, and neuroscience-based learning enhancements may redefine the way knowledge is acquired. The hope embedded in these innovations is the promise of a world where education is accessible to all, transcending geographic and economic barriers. Learning will no longer be constrained by rigid curricula, as adaptive AI tailors instruction to individual strengths, interests, and career trajectories.

Yet fear remains a persistent shadow. As technology redefines education, the traditional markers of achievement—degrees, standardized testing, and even schooling- may become obsolete. Will students be prepared for a future where knowledge is fluid and rapidly evolving? Moreover, a few tech giants' monopolization of AI-driven education could create new forms of intellectual inequality, where access to the best learning tools is restricted to the privileged.

Despite these challenges, hope endures through collaborative learning networks that empower individuals to self-direct their education. The rise of open-access knowledge hubs and decentralized credentialing systems may counteract monopolization, ensuring that education remains a force for social mobility rather than division.

Religion: The Convergence of Faith and "*Digital Divinity*" (Youvan,2023).

Religion, often regarded as humanity's response to existential hope and fear, will not remain untouched by the sweeping technological and economic changes. The digital age has already witnessed the rise of virtual congregations, AI-generated religious texts, and cyber-rituals. In the coming decades, faith traditions may continue evolving into hybrid systems where human and artificial spirituality intertwine.

Hope manifests in the possibility of interfaith unity and increased spiritual accessibility. Virtual reality temples, AI-driven spiritual advisors, and immersive religious experiences could bring people together across traditional divides, fostering a more profound sense of global community. Transhumanist philosophies may blend with religious thought, presenting new avenues for exploring consciousness, the afterlife, and the divine.

However, fear arises from the potential dilution of religious traditions and the ethical dilemmas posed by AI-driven spirituality.

As digital prophets emerge, who will determine religious authority? Will faith become yet another algorithm-driven commodity? Some may resist these changes, leading to tensions between traditionalists and technologists, between those who seek solace in ancient rituals and those who embrace "digital divinity."

The Future of Hope and Fear in Economics, Education, and Religion

In the future, the concepts of hope and fear will continue to shape the trajectories of society, particularly in the interconnected domains of economics, education, and religion. These human emotions, abstract, have profound impacts on how we envision the world, our future, and the choices we make. As technological advancements, economic shifts, and evolving cultural landscapes alter the structure of these domains, hope and fear will evolve in complex ways. This essay speculates on how these emotions might play out across the three domains, reflecting on their changing role in shaping our collective experience.

I. Economics: The Duality of Hope and Fear in an Age of Technological Transformation

Economics, the system that governs the distribution of resources and wealth, is intricately linked with human emotions. Historically, hope has been a driving force for innovation, while fear often motivates stability and risk aversion. In the future, as automation, artificial intelligence (AI), and other technological advancements continue to transform the global economy, these emotions will be intensified and manifested in new ways.

A. The Hope for Technological Prosperity

In the future, hope will be primarily driven by technological progress. Automation and AI are poised to revolutionize industries, increasing productivity and creating new markets. The hope for economic prosperity may come from the belief that these technological advancements will result in a more efficient, equitable, and sustainable global economy. For instance, the advent of fully automated factories or self-driving vehicles could lead to economic abundance, with goods becoming cheaper and more widely available.

Furthermore, hope will also be tied to the promise of new forms of wealth and economic participation. The rise of the gig economy, the expansion of digital currencies like cryptocurrencies, and the decentralization of financial systems may give individuals more control over their economic destinies. People could hope for greater opportunities for entrepreneurship, self-sufficiency, and financial independence, even in the face of an increasingly automated workforce.

B. The Fear of Economic Disruption

However, these hopes may be tempered by profound fears. Automating jobs and increasing AI-driven processes could lead to massive economic displacement. As technology evolves exponentially, many workers in the manufacturing and healthcare sectors could face obsolescence. This could create a pervasive fear of job insecurity, economic inequality, and social unrest as substantial portions of the population struggle to adapt to the new economic realities.

Fear of economic collapse could also emerge from the unanticipated consequences of technological advancements. Overreliance on digital economies, for instance, could make societies vulnerable to cyber-attacks, hacking, and data breaches. Similarly, the economic impacts of climate change and the ecological crisis could lead to fears about the future stability of markets, especially in resource-dependent economies.

C. A Future of Economic Polarization?

The dualistic environment of hope and fear shapes the economic landscape as individuals and societies oscillate between optimism for growth and apprehension about instability. Hope drives innovation, investment, and the pursuit of opportunities, while fear often leads to caution, risk aversion, and protective measures. This interplay becomes particularly pronounced during periods of economic uncertainty, where disparities in wealth and access to resources exacerbate tensions. Increasing polarization in economics is evident as wealth gaps widen, with affluent groups benefiting from technological advancements and globalization while marginalized communities face stagnation or decline. This division fuels social and political unrest as differing priorities and values clash over resource allocation and policy decisions. The future of economics may be marked by heightened polarization as technological disruptions and climate challenges further deepen inequalities. Addressing this duality requires a balanced approach that fosters equitable growth while mitigating fears through inclusive policies and collaborative efforts. This divide could generate societal instability and lead to new forms of economic resistance, such as universal basic income (UBI) movements, labor strikes, or even the rise of alternative economic systems that challenge the capitalist status quo.

II. Education: Navigating Hope and Fear in the Learning Landscape

Education, the process of transmitting knowledge, skills, and values, is fundamental in shaping future generations' perceptions of hope and fear. As technology disrupts traditional education models and society increasingly places value on adaptability and lifelong learning, the role of hope and fear in education will be more pronounced.

A. The Hope for Access to Knowledge and Empowerment

In the future, hope will center on the democratization of knowledge. Online platforms, virtual classrooms, and AI-driven learning tools will make education more accessible than ever before. People can learn at their own pace, on their terms, with resources at their fingertips. This promises to break down barriers to education and empower individuals, regardless of geographic location or socioeconomic status.

Hope in education will also be associated with personal growth and transformation. The ability to continually learn and adapt will become increasingly vital in an economy where jobs change rapidly and the skills of today may be obsolete tomorrow. There will be a renewed hope for education as the key to upward mobility, self-empowerment, and social advancement.

B. The Fear of Educational Inequality and Inadequacy

Despite the optimism surrounding access to education, fears about the growing inequality in educational opportunities will persist. While some will benefit from advanced technologies and personalized learning environments, others will struggle to keep up, particularly in poor regions. The digital divide, which already limits access to the internet and digital devices in many parts of the world, could exacerbate educational inequality, leaving marginalized populations further behind.

Moreover, as the demand for specialized knowledge increases, there may be a fear of educational inadequacy. Traditional educational institutions may struggle to keep pace with the speed of technological change. Fears of obsolescence, especially among older generations, could lead to anxiety about whether educational systems will adequately prepare students for an ever-changing world.

C. A Global Education System in Crisis?

As global inequality continues to widen, the future may witness the emergence of a two-tiered education system that starkly divides access to resources and opportunities. The privileged few, often from affluent backgrounds, will likely benefit from cutting-edge technologies, personalized learning experiences, and innovative curricula tailored to foster creativity and critical thinking. These students may have access to advanced tools such as artificial intelligence, virtual reality, and specialized mentorship programs that prepare them for leadership roles in a rapidly evolving world. In contrast, the majority, particularly those from underprivileged communities, may be confined to outdated educational models that lack the resources to adapt to modern demands. These schools may struggle with overcrowded classrooms, insufficient funding, and a curriculum that fails to address the skills needed for the future workforce. This disparity perpetuates a cycle of inequality, as students from disadvantaged backgrounds are left ill-equipped to compete in a globalized economy. The two-tiered system risks deepening social divides as education becomes a determinant of economic and social mobility. Addressing this challenge requires systemic reforms, equitable resource distribution, and policies that prioritize inclusive education for all. Without such efforts, the gap between the privileged and the marginalized may continue to grow, undermining the potential for a more equitable and just society.

III. Religion: Hope and Fear in a Changing Spiritual Landscape

Religion, as a domain deeply intertwined with existential questions, is uniquely positioned to navigate the evolving dynamics of hope and fear. As science and technology advance, they challenge traditional religious narratives by offering alternative explanations for phenomena once attributed to divine forces. This progress may lead some individuals to question the relevance of religion in a world increasingly driven by empirical evidence and innovation. However, the emotional forces of hope and fear remain central to the human experience, ensuring that religion continues to play a significant role in shaping spiritual lives. Hope, often rooted in the promise of salvation or a better future, provides comfort and motivation. At the same time, fear, tied to uncertainties about morality and the afterlife, compels introspection and adherence to ethical principles. Together, these emotions drive individuals to seek meaning and connection, which religion has historically provided through rituals, community, and moral guidance. In this dualistic environment, religion may adapt by integrating scientific insights and addressing contemporary challenges, ensuring its relevance in a rapidly changing world. Ultimately, the interplay of hope and fear will continue to influence how individuals engage with spirituality, shaping the evolution of religious thought and practice.

A. The Hope for Meaning and Salvation in an Uncertain World

Religion, with its promise of salvation, purpose, and an afterlife, will provide a source of hope for many. In an increasingly secular and technology-driven world, people will continue to seek meaning and purpose in their lives. The quest for meaning and purpose remains a fundamental aspect of human existence. As scientific advancements and technological innovations reshape societies, traditional religious frameworks may face challenges in maintaining their relevance. However, the emotional and existential needs of individuals persist, driving many to seek solace and guidance in religion. With its enduring promise of salvation, a sense of purpose, and the hope of an afterlife, religion continues to offer a profound source of comfort and inspiration. It provides a moral compass and a community of shared beliefs, fostering resilience in the face of uncertainty. Even as secular ideologies gain prominence, the spiritual dimensions of hope and fear remain deeply ingrained in human psychology, ensuring that religion retains its significance. By adapting to contemporary challenges and integrating new perspectives, religious institutions can continue to address the evolving needs of individuals. Ultimately, the interplay between secular progress and spiritual longing highlights the enduring role of religion in shaping human lives and aspirations. As social and economic upheaval intensifies, the spiritual solace provided by religion will offer individuals a sense of stability and connection to something greater than themselves.

Religious institutions may evolve in the future, adapting to new technological realities by offering virtual worship services, AI-guided spiritual counseling, and online religious education. For those seeking a sense of community, religious organizations may become hubs for support and solidarity in a world that feels increasingly fragmented.

B. The Fear of Disillusionment and Loss of Faith

At the same time, fear will be a prominent emotion within the religious sphere. As scientific discoveries continue to challenge traditional religious beliefs, many individuals may experience fear and anxiety over the loss of meaning and purpose in their lives. The rise of secularism and the decline of religious participation in many parts of the world could lead to widespread fear of spiritual emptiness and existential crisis.

Additionally, as religions evolve in response to modernity, there may be fears of distortion and corruption of traditional spiritual teachings. As new religious movements emerge and challenge established doctrines, adherents may fear the loss of authenticity and the erosion of their spiritual identity.

C. A Future of Spiritual Evolution or Fragmentation?

In the future, religion may either undergo significant evolution or fragment into various subcultures. Some may embrace a more flexible, inclusive approach to spirituality, while others may retreat into more fundamental or traditional beliefs in response to societal pressures. The future of religion will be shaped by the balance between hope for a transcendent reality and fear of losing one's faith in a rapidly changing world.

The future of hope and fear in economics, education, and religion will be marked by profound transformations driven by technological progress, societal change, and existential challenges. These emotions will continue to guide individuals and societies, shaping how people navigate their futures. The hope for economic prosperity, educational empowerment, and spiritual meaning will inspire new possibilities, while the fear of economic inequality, educational inadequacy, and spiritual disillusionment will continue to challenge progress. The interplay between hope and fear will undoubtedly create a complex and dynamic future, one in which humanity must find ways to adapt, innovate, and reconcile these powerful emotions to build a more sustainable and equitable world.

A Balancing Act of Hope and Fear

As humanity ventures further into the 21st century, the interplay between hope and fear in economics, education, and religion will shape the very fabric of society. The forces of technological advancement, decentralization, and globalization hold both immense promise and profound peril. Whether hope triumphs over fear will depend on humanity's ability to harness innovation for collective good while mitigating the risks that accompany rapid change.

In this speculative future, we may find ourselves standing on the precipice of unprecedented prosperity or unforeseen chaos. The choices we make today—about how we shape economies, educate future generations, and define spiritual fulfillment—will determine whether hope remains the guiding force of civilization or whether fear takes the reins. One thing, however, is certain: the future will be neither static nor predictable but rather a dynamic interplay of our greatest aspirations and deepest anxieties.

Chapter 8

THE INTERCONNECTEDNESS OF RELIGION AND POLITICS

A Historical Perspective

Religion and politics have been deeply intertwined throughout history, significantly shaping societal norms, governance, and collective identities. This interconnectedness influences policy-making, legitimacy, social cohesion, and global relations.

One of the most evident aspects of the interplay between religion and politics is the moral framework that religion provides to political systems. Religions often advocate values such as justice, equality, and compassion, which become integral to political ideologies (Smith, 2020). In democratic societies, religious beliefs can influence public opinion and policy development, as legislators often craft laws aligned with widely accepted moral and ethical standards. For example, social justice movements advocating for equality have drawn inspiration from religious principles.

Additionally, political leaders frequently use religious narratives to legitimize their authority and gain public support. Historical examples such as the divine right of kings illustrate how rulers aligned themselves with religious institutions to reinforce their political power (Jones & Wilkinson, 2018). Similarly, modern political campaigns often incorporate religious rhetoric to resonate with constituents. This dynamic highlights how religion serves as a tool for achieving political legitimacy. In ancient civilizations, religion was often inseparable from politics. Rulers claimed divine authority and religious institutions played a crucial role in governance. In Ancient Egypt, pharaohs were regarded as gods and religious beliefs legitimized their rule. Similarly, in Mesopotamian societies, kings were seen as chosen by the gods to maintain cosmic order. The intertwining of religion and politics was also evident in Ancient China, where the Mandate of Heaven justified the legitimacy of emperors. The idea that rulers governed with divine approval reinforced political stability and justified rebellion when rulers were deemed to have lost divine favor.

In these early societies, religious doctrines were instrumental in shaping laws, societal structures, and political legitimacy. Religion and politics also play a dual role in fostering social cohesion and division. Shared religious values can unite communities, creating a sense of collective identity and purpose. In times of crisis, religious institutions often serve as sanctuaries that provide emotional and material support to individuals, reinforcing social bonds (Ahmed, 2019). However, the politicization of religious identities can exacerbate divisions, leading to sectarian conflicts and discrimination. When political leaders exploit religious differences for gain, societal harmony is jeopardized. On a global scale, the relationship between religion and politics varies widely depending on cultural, historical, and social contexts. In secular states, religion's role in politics is limited, emphasizing separation to maintain neutrality. In contrast, theocratic states integrate religion into their political systems, where governance and religious doctrines are inseparable (Taylor, 2021). Both approaches highlight the diverse ways societies navigate this complex relationship.

Religion and politics are deeply interconnected, shaping societal structures and norms. While religion provides moral frameworks and political leaders seek legitimacy through religious narratives, the interplay between the two can both unite and divide societies. Understanding this relationship is essential for comprehending the foundations of governance, culture, and collective identity. The relationship between religion and politics has shaped societies and civilizations throughout history. Religion has often influenced political ideologies, policies, and governance, while politics has, in turn, affected religious practices and institutions. This dynamic interplay has led to significant historical events, conflicts, and transformations. By examining various historical periods and societies, this essay explores how religion and politics have interacted and influenced each other.

The medieval period saw a complex relationship between religious and political authorities. In Europe, the Catholic Church wielded immense power, influencing monarchs and shaping political decisions. The investiture controversy of the 11th and 12th centuries, a conflict between Pope Gregory VII and Emperor Henry IV of the Holy Roman Empire, highlighted tensions over whether secular rulers or the Pope had the authority to appoint bishops.

On the other hand, the Byzantine Empire embraced caesaropapism, where the emperor held supreme authority over the state and the Church. This fusion of religious and political power strengthened imperial rule and led to conflicts, such as the Great Schism of 1054, which divided Christianity into Roman Catholicism and Eastern Orthodoxy.

In the Islamic world, the Caliphate represented a system where religious and political leadership were combined. The early Caliphs were both spiritual and political leaders and Sharia law influenced governance. However, different interpretations of Islam led to divisions, as seen in the Sunni-Shia split, which had profound political implications.

The Reformation and Religious Wars

The Protestant Reformation of the 16th century significantly altered the relationship between religion and politics in Europe. Martin Luther's challenge to the Catholic Church's authority led to the fragmentation of Christendom and the emergence of Protestant states. Political leaders, such as Henry VIII of England, used religious reformation to consolidate their own power by breaking away from papal authority and establishing national churches.

The Reformation also sparked violent conflicts, such as the Thirty Years' War (1618-1648), which was both a religious and political struggle involving Protestant and Catholic states. The Peace of Westphalia, which ended the war, established the principle of state sovereignty, diminishing the political power of the Catholic Church and paving the way for the modern secular state.

The Enlightenment period brought significant shifts in the relationship between religion and politics. Thinkers such as Voltaire, John Locke, and Jean-Jacques Rousseau advocated for the separation of church and state, religious tolerance, and individual rights. The American and French Revolutions were influenced by these ideas, leading to the establishment of secular governments where religious authority was distinct from political power.

The U.S. Constitution enshrined the principle of religious freedom, ensuring that the government would neither establish a national religion nor interfere with religious practices. Similarly, the French Revolution saw the radical secularization of the state, leading to the confiscation of Church properties and the promotion of a secular civic religion.

Modern Implications and Continuing Interactions

Despite the rise of secularism, religion continues to play a role in modern politics. In many nations, religious beliefs influence political debates on issues such as abortion, same-sex marriage, and education. As stated previously in this chapter, political leaders often appeal to religious values to garner support, while religious institutions advocate for policies aligned with their beliefs.

In some regions, the fusion of religion and politics remains strong. In Iran, the Islamic Republic is governed by a theocratic system where religious leaders hold significant political power. In contrast, countries such as Türkiye have experienced tensions between secular and religious factions, reflecting ongoing debates about the role of religion in governance.

The interplay between religion and politics has been a defining feature of human history, shaping societies, laws, and governance. From ancient theocracies to modern secular states, this relationship has evolved in response to social, intellectual, and political changes. While secularism has gained prominence, religion remains a potent force in political discourse. Understanding this historical interplay provides insight into contemporary political and religious dynamics, highlighting the enduring influence of faith in the public sphere.

Chapter 9

FAITH AND KNOWLEDGE IN EDUCATION

The connection between religion and secular education has been a defining feature of academic discourse for centuries. As societies have evolved, the tension between faith and reason has shaped policies, curricula, and individual experiences in profound ways. From the early days of religiously affiliated schools to the rise of secular public education, the balancing act between maintaining religious identity and fostering a comprehensive, inclusive learning environment has remained a central concern. This historical narrative discusses the role of religion in secular education, the struggle to balance faith and reason in the curriculum, and personal stories of individuals navigating religious identity in educational settings.

A Historical Perspective of Religion in Secular Education

The intersection of faith and knowledge in education is a profound area of exploration, as it shapes how individuals perceive the world, develop critical thinking, and integrate moral values into their learning. A significant theme emergent in the discussion of faith-based education is the importance of holistic development. Faith-based educational leaders often emphasize the development of the whole person—intellectually, morally, and spiritually. This approach fosters a balanced perspective, encouraging students to engage with both empirical knowledge and ethical considerations (Smith, 2023). Moral and Ethical frameworks are also noted as important dimensions of education. Faith traditions provide moral frameworks that can guide educational practices and decision-making. For instance, principles such as compassion, justice, and integrity are often integrated into curricula, influencing how students approach societal challenges (Jones & Taylor, 2022). The pedagogy of faith-based learning often stresses the importance of Critical Thinking and Dialogue. The integration of faith and knowledge encourages critical thinking by prompting students to question and reflect on their beliefs in light of new information. This dialogue between faith and reason can lead to a deeper understanding of complex issues (Ahmed, 2021). Taylor (2024) hypothesizes that Cultural and Social Impact in Faith-based education causes reflection on the cultural and social contexts in which it operates, promoting community engagement and social responsibility. This intersection can help students connect their learning to real-world applications, fostering a sense of purpose and service.

The roots of education in many societies are deeply intertwined with religious instruction. In medieval Europe, the Church played a dominant role in schooling, with institutions like the University of Paris and Oxford University established initially to train clergy. The Renaissance and Enlightenment eras introduced a shift toward reason and scientific inquiry, challenging the absolute authority of religious doctrine in academic settings.

In the United States, early education was primarily religious, with the Puritans establishing schools to ensure biblical literacy. However, as the nation expanded, a growing emphasis on religious pluralism led to the development of public schools that sought to provide a neutral, secular education. The 19th-century Common School Movement, led by Horace Mann, advocated for a curriculum free from denominational influences, aiming to create a shared civic identity among diverse religious communities.

Despite efforts to secure education, religious debates persisted. Landmark Supreme Court cases, such as Engel v. Vitale (1962), which ruled against state-sponsored prayer in schools, and Lemon v. Kurtzman (1971), which established the "Lemon Test" for determining the constitutionality of religious involvement in public education, further solidified the separation of church and state. Nevertheless, questions about the place of religion in education continue to spark debates in contemporary society.

Balancing Faith and Reason in the Curriculum

As secular education evolved, the challenge of integrating religious perspectives while upholding academic integrity became increasingly complex. The debate over teaching evolution versus creationism is a prime example of this tension. In the early 20th century, the Scopes Trial (1925) underscored the cultural divide between religious fundamentalists and advocates of scientific reasoning. While public schools embraced evolutionary theory, some religious communities continued to push for alternative explanations that aligned with their beliefs.

Beyond science, subjects like literature, philosophy, and history often intersect with religious themes. Educators have grappled with how to teach religious topics in an objective, non-proselytizing manner. Comparative religion courses emerged to educate students about various faith traditions without endorsing any belief system. These courses aim to promote religious literacy and foster mutual understanding in increasingly diverse classrooms.

Another critical issue is the accommodation of religious practices within secular schools. From allowing students to wear religious attire to providing prayer spaces and dietary accommodations, schools have had to navigate policies that respect religious expression while maintaining a neutral learning environment. Implementing such accommodations reflects an ongoing effort to balance personal faith with the principles of secular education.

Navigating Religious Identity in Educational Settings

For many students and educators, reconciling religious identity with secular education presents both challenges and opportunities. One such story is that of Aisha, a Muslim student attending a public high school in a non-Muslim community. She faced difficulties finding spaces to pray during school hours and often felt isolated when her peers did not understand her religious observances. However, through advocacy and dialogue with school administrators, Aisha successfully helped establish a designated prayer space, fostering greater inclusivity for students of all faiths.

Similarly, Michael, a devout Christian science teacher, struggled with reconciling his faith with the curriculum he was required to teach. While he personally believed in intelligent design, he adhered to the mandated scientific standards, emphasizing critical thinking and allowing students to explore different viewpoints. His experience highlights the nuanced approach many educators take to uphold both academic integrity and personal beliefs.

On the other hand, Sarah, an atheist student in a religiously conservative town, encountered resistance when she questioned the religious perspectives presented in her school. Facing social ostracization, she found solace in a philosophy class that encouraged open dialogue on existential and moral questions. Her journey underscores the importance of academic spaces that encourage diverse perspectives and critical engagement with religious and secular ideas alike.

The relationship between religion and secular education continues to evolve, shaped by historical shifts, legal battles, and individual experiences. While tensions persist, efforts to balance faith and reason in the curriculum demonstrate a commitment to fostering both academic rigor and respect for diverse religious identities. Personal narratives reveal the real-world implications of these debates, illustrating the ways individuals navigate their beliefs in educational settings. Moving forward, a nuanced approach that encourages dialogue, understanding, and inclusivity will be essential in bridging the divide between faith and knowledge in the modern classroom.

Chapter 10

THE POWER OF HOPE IN TRANSFORMING EDUCATION

Hope is a fundamental force in education, driving both students and educators to envision and strive for a better future. It serves as a catalyst for resilience, motivation, and innovation, enabling individuals to overcome challenges and achieve their goals. In educational settings, hope fosters a positive learning environment, encourages perseverance, and inspires creativity. By integrating hope into pedagogical practices, educators can empower students to develop a growth mindset and embrace lifelong learning. The power of hope in education lies in its ability to promote resilience and motivation. When students face academic or personal challenges, hope provides them with the strength to persevere and seek solutions. Research indicates that hopeful students are more likely to set ambitious goals and work diligently to achieve them (Webb, 2019). This mindset not only enhances academic performance but also prepares students to navigate life's uncertainties with confidence.

Moreover, hope plays a crucial role in fostering a positive learning environment. Educators who cultivate hope in their classrooms create spaces where students feel valued, supported, and inspired. This approach encourages collaboration, creativity, and critical thinking, enabling students to explore new ideas and perspectives (Green, 2020). By emphasizing the importance of hope, educators can help students develop a sense of purpose and direction. Hope is an essential element in education, serving as a driving force that empowers students to overcome challenges and achieve their goals. Schools and educational institutions worldwide have recognized the importance of instilling hope in students, leading to the creation of various initiatives and programs aimed at fostering resilience, motivation, and academic success. This chapter details key programs that promote hope in education, highlights the success stories of individuals and communities, and examines the pivotal role of educators in nurturing an environment of optimism and possibility.

Numerous programs have been developed to cultivate hope in students, particularly those from underprivileged backgrounds. These initiatives aim to provide students with academic support, mentorship, and opportunities that enable them to envision a brighter future.

One such initiative is the **Gaining Early Awareness and Readiness for Undergraduate Programs (GEAR UP)**, a federally funded initiative in the United States designed to increase college readiness among low-income students. Through mentorship, academic counseling, and financial aid information, GEAR UP has helped thousands of students access higher education, reinforcing the belief that academic success is attainable despite socioeconomic challenges.

Another prominent program is the **Khan Academy**, a free online educational platform that offers resources across multiple subjects. By providing students with self-paced learning tools, Khan Academy fosters a sense of control over their academic progress, instilling confidence and hopes that they can improve their skills and achieve their goals.

Additionally, the **Posse Foundation** is a scholarship program that identifies talented students from diverse backgrounds and provides them with full-tuition scholarships and leadership training. The foundation's emphasis on peer support and mentorship has helped students develop resilience and a powerful sense of belonging, increasing college retention and graduation rates.

Educational programs that prioritize hope and empowerment have transformed the lives of countless individuals and communities. One inspiring example is that of **Dr. Victor Rios**, a former gang member who, with the support of an encouraging teacher, turned his life around to become a professor and advocate for at-risk youth. His journey highlights the transformative power of mentorship and positive reinforcement in fostering hope.

Also, the community of **Camden, New Jersey**, once plagued by low graduation rates and high crime levels, has witnessed significant improvement due to educational initiatives such as the **Promise Neighborhoods Program**. This initiative offers comprehensive support, including early childhood education, college readiness programs, and community engagement, creating a pathway to success for students who once had limited opportunities.

In South Africa, the **LEAP Science and Maths Schools** provide quality education to students from disadvantaged communities, focusing on holistic development, mentorship, and leadership training. Many graduates of LEAP schools have gone on to pursue higher education and contribute positively to their communities, demonstrating the impact of hope-driven education.

The Role of Educators in Nurturing Hope

Hope is a transformative force in education, empowering students to envision a brighter future and overcome challenges. In today's rapidly changing world, the role of educators in nurturing hope has become increasingly urgent. By fostering resilience, motivation, and a sense of purpose, educators can help students navigate uncertainties and embrace lifelong learning. This paper explores the critical role educators play in cultivating hope and its profound impact on educational outcomes. Educators are uniquely positioned to nurture hope by creating supportive and inclusive learning environments. Research indicates that students who feel valued and supported by their teachers are more likely to develop a hopeful mindset and achieve academic success (Sulimani-Aidan et al., 2019). By fostering a sense of belonging, educators can help students build confidence and resilience, enabling them to overcome obstacles and pursue their goals.

Moreover, educators inspire hope by encouraging students to set ambitious goals and work toward them. Through mentorship and guidance, educators help students identify pathways to success and develop the skills needed to navigate life's challenges. This process not only enhances academic performance but also equips students with the tools to envision and strive for a better future (Freire, 2014). By instilling hope, educators empower students to become agents of change in their communities.

One essential aspect of fostering hope in students is **growth mindset encouragement** (Dweck, 2017). Educators who emphasize the value of effort, learning from mistakes, and persistence help students develop resilience. Research by psychologist Carol Dweck has shown that students who believe in their ability to grow and improve are more likely to overcome academic and personal challenges.

Moreover, educators serve as **mentors and role models**, offering guidance and emotional support to students facing adversity. Programs like **Big Brothers Big Sisters** have demonstrated that mentorship significantly increases students' confidence and academic engagement, reinforcing their belief in a hopeful future.

Another critical element is integrating **social-emotional learning (SEL)** into school curricula. SEL programs teach students essential skills such as self-awareness, empathy, and emotional regulation, equipping them with the tools to navigate obstacles confidently and optimistically. Schools implementing SEL frameworks report higher student engagement, lower dropout rates, and improved overall well-being.

Hope is a powerful educational force, enabling students to overcome obstacles and envision a brighter future. Through initiatives like GEAR UP, Khan Academy, and the Posse Foundation, students from disadvantaged backgrounds are given the resources and support needed to succeed. Success stories such as that of Dr. Victor Rios and transformative community programs like the Promise Neighborhoods Program further illustrate the profound impact of fostering hope in educational settings. Educators are pivotal in nurturing hope by encouraging growth mindsets, serving as mentors, and integrating social-emotional learning. By continuing to invest in hope-driven education, society can create lasting changes and empower future generations to reach their fullest potential.

Finally, hope inspires innovation and progress in education. It challenges educators and policymakers to envision transformative changes that address societal needs and promote equity. By integrating hope into educational practices, institutions can empower students to become agents of change and contribute to the betterment of their communities (Ahmed, 2021). The power of hope in education is essential for fostering resilience, motivation, and innovation. By embracing hope as a guiding principle, educators can create transformative learning experiences that empower students to achieve their full potential and make meaningful contributions to society. The urgency of nurturing hope in education is further underscored by the need to address societal challenges and promote equity. Educators play a crucial role in inspiring students to confront social injustices and contribute to the betterment of society. By integrating hope into their teaching practices, educators can foster a sense of purpose and direction, encouraging students to make meaningful contributions to the world (Albarracin, 2024).

The role of educators in nurturing hope is essential for fostering resilience, motivation, and a sense of purpose in students. By embracing hope as a guiding principle, educators can create transformative learning experiences that empower students to achieve their full potential and make a positive impact on society. The urgency of nurturing hope in education is further underscored by the need to address societal challenges and promote equity. Educators play a crucial role in inspiring students to confront social injustices and contribute to the betterment of society. By integrating hope into their teaching practices, educators can foster a sense of purpose and direction, encouraging students to make meaningful contributions to the world (Albarracin, 2024). The role of educators in nurturing hope is essential for fostering resilience, motivation, and a sense of purpose in students. By embracing hope as a guiding principle, educators can create transformative learning experiences that empower students to achieve their full potential and make a positive impact on society.

Chapter 11

CONFRONTING FEAR IN THE POLITICAL ARENA

Fear is a powerful force in politics, often used to influence public opinion, shape policies, and mobilize support. While fear can be an effective motivator, it also has the potential to exacerbate divisions, perpetuate biases, and hinder progress. Confronting fear in politics is essential for fostering a more inclusive and equitable society. Moreover, addressing the politics of fear has profound implications for education, as it shapes how individuals learn, think critically, and engage with diverse perspectives.

In the realm of politics, fear is frequently employed as a tool to sway opinions and drive behavior. Politicians often use fear-based messaging to highlight risks and threats, thereby encouraging individuals to align with their agendas. However, this strategy can lead to polarization and the marginalization of certain groups. Research suggests that confronting fear requires fostering dialogue, promoting empathy, and encouraging critical thinking (Albarracin, 2024). By addressing fear, political systems can move toward more constructive and inclusive approaches.

The implications of fear in politics extend to education, where it influences curriculum design, pedagogy, and student engagement. Fear-based narratives can create boundaries between "us" and "them," perpetuating biases and limiting opportunities for critical inquiry (Zembylas, 2009). Educators play a crucial role in challenging these narratives by fostering environments that encourage open dialogue, critical thinking, and empathy. By confronting fear in educational settings, students can develop the skills needed to navigate complex social and political landscapes. Educators are essential in nurturing hope, as they inspire resilience, motivation, and a sense of purpose in their students. By fostering hope, educators empower students to achieve their full potential and make meaningful contributions to society.

The Role of Activism and Grassroots Movements

Activism and grassroots movements have effectively countered political fear. By mobilizing ordinary citizens, these movements create collective resistance that is difficult for oppressive systems to dismantle. One of the most significant examples is the American civil rights movement of the 1950s and 1960s. Activists like Rosa Parks, Martin Luther King Jr., and John Lewis led nonviolent protests, sit-ins, and voter registration drives, challenging segregation and systemic racism despite violent backlash.

Similarly, grassroots movements have played crucial roles in dismantling authoritarian regimes. The Solidarity movement in Poland, led by Lech Wałęsa, was instrumental in resisting communist rule in the 1980s. Despite government crackdowns, the movement persisted, contributing to the fall of communism in Eastern Europe.

In South Africa, the anti-apartheid movement demonstrated the power of sustained grassroots activism. Organizations such as the African National Congress (ANC) mobilized both domestic and international support, leading to the eventual dismantling of apartheid and the election of Nelson Mandela as the country's first Black president in 1994. More recently, movements like the Arab Spring uprisings of 2011 and the pro-democracy protests in Hong Kong have illustrated the continuing relevance of grassroots activism in challenging political oppression. While these movements have met with varying degrees of success, they highlight the courage of ordinary citizens in confronting fear and demanding change.

History has witnessed the rise of leaders who have countered political fear with messages of hope and unity. These leaders have demonstrated moral courage, resilience, and the ability to mobilize people against oppression. One of the most iconic leaders in this regard was former President Franklin D. Roosevelt, who, during his inaugural address in 1933, famously stated, "The only thing we have to fear is fear itself." Amidst the Great Depression, Roosevelt's leadership reassured the American public and implemented policies under the New Deal that provided economic relief and recovery, alleviating widespread fear and uncertainty.

Martin Luther King Jr. is another example of a leader who inspired hope in the face of fear. His philosophy of nonviolent resistance, articulated in speeches such as "I Have a Dream" and "Letter from Birmingham Jail," encouraged African Americans to stand against segregation and discrimination without succumbing to fear or hatred. Nelson Mandela's leadership in South Africa is another compelling example. After spending 27 years in prison, Mandela emerged as a symbol of reconciliation rather than revenge. His leadership in guiding South Africa through a peaceful transition from apartheid to democracy demonstrated the power of hope in overcoming deep-seated political fear and division.

In the modern era, figures such as Malala Yousafzai have continued this legacy. Malala Yousafzai's journey is a testament to resilience and unwavering commitment to justice. After surviving a life-threatening attack by the Taliban in 2012, she transformed her personal tragedy into a global movement for girls' education and human rights. Her advocacy began at a young age when she courageously spoke out against the Taliban's ban on girls' education in Pakistan's Swat Valley. Through her speeches, writings, and activism, Malala has become a symbol of hope for marginalized communities worldwide, emphasizing the transformative power of education. Her efforts have earned her numerous accolades, including the Nobel Peace Prize, making her the youngest laureate in history. Malala's work extends beyond rhetoric; she has established the Malala Fund, which supports education initiatives in regions affected by conflict and poverty. Her story inspires millions to challenge oppression and strive for equality, proving that even in the face of adversity, one voice can ignite change on a global scale. Malala's legacy continues to grow, reminding the world of the importance of education as a fundamental human right. Malala has remained a steadfast advocate for girls' education and human rights, inspiring hope among millions around the world.

Political fear has long been used as a tool of oppression, but history provides numerous examples of how it can be mitigated and overcome. Strategies such as independent media, legal protections, education, and civic engagement have effectively countered fear-based governance. Activism and grassroots movements have played essential roles in mobilizing collective resistance, while visionary leaders have inspired hope and resilience. By learning from these historical lessons, societies can continue to resist fear and build more just and democratic futures.

Confronting fear in politics is vital for promoting inclusivity and equity. Its implications for education highlight the importance of fostering critical thinking, empathy, and dialogue. By addressing fear in both political and educational contexts, society can work toward a more just and equitable future.

Chapter 12

NAVIGATING RELIGIOUS UNCERTAINTY:

Two Case Studies and Relevant Research Studies

Religious uncertainty is a complex phenomenon that individuals and communities experience in diverse ways, often shaped by personal beliefs, cultural influences, and social interactions. Qualitative case studies provide an invaluable framework for exploring these uncertainties, as they offer deep, context-rich analyses of how individuals navigate shifts in faith, religious skepticism, and doctrinal tensions. By employing methods such as in-depth interviews and ethnographic observations, researchers can capture the cognitive and emotional strategies that individuals use to reconcile doubt, reinterpret traditions, or negotiate conflicting ideologies. Case studies allow readers to engage with the lived experiences of those grappling with religious uncertainty, fostering a nuanced understanding of coping mechanisms such as reflective dialogue, community support, and spiritual reconfiguration (Creswell & Poth, 2018). These insights contribute to broader discussions on religious resilience, identity formation, and the ways in which uncertainty can lead to transformation rather than mere dissonance (Smith & Osborn, 2015). The following two case studies depict uncertainty in the spiritual arena. They tell the stories of Edgar and Maria, each struggling with their own uncertainty.

The first time Edgar questioned his faith was when he was seventeen. He had grown up in a devout household where Sunday services were non-negotiable; everyone was expected to attend. Prayers were whispered before every meal; scripture guided every decision. However, one evening, as he lay awake staring at the ceiling, a question gnawed at him: What if none of it was true?

Edgar's doubt began with small cracks—an unanswered prayer, a contradiction in a sermon, a tragedy that seemed unfair. As he entered college, those cracks deepened. His philosophy professor challenged him to consider alternate worldviews, and his friends questioned why he held onto beliefs he could not fully explain. The once-solid foundation of his faith felt unsteady, and for the first time in his life, he felt alone.

The first chapter of the Book of James in the New Testament is profoundly significant. It focuses on key themes such as finding joy amid trials, seeking wisdom from God, and understanding the destructive potential of uncertainty. It emphasizes the value of perseverance during challenges, encouraging believers to trust in divine guidance. Furthermore, this chapter underscores the need to anchor faith in everyday actions, illustrating how spiritual conviction can provide strength and purpose even in difficult times. A study of the Book of James can help individuals address uncertainties in their faith, examine church doctrines critically, and ensure that personal knowledge and commitment to faith are consistent. The need to be prayerful and intentional in study is genuine for all faiths.

In another case, Maria Mitchell, a Unitarian astronomer, (Alary &Rooney, 2023), confronts questions that arise in her academic research. Maria's journey reflects the value of seeking and questioning truth in science and spirituality. This narrative demonstrates how doubt and uncertainty can contribute to a more profound comprehension of the subject being questioned.

Across the country, Maria was going through a struggle similar to that of seventeen-year-old Edgar. Raised in a deeply religious Catholic family, she had always taken comfort in the rituals and traditions of the church. Nevertheless, when her younger brother was diagnosed with a terminal illness, her faith was shaken. She prayed relentlessly, begging for a miracle that never came. After his passing, she found herself resenting the very beliefs that had once given her solace. Why would a loving God allow such suffering? Why did her prayers go unanswered?

Both Edgar and Maria withdrew from their religious communities, unsure of where they belonged. Edgar stopped attending church, finding it hypocritical to participate in something he no longer felt connected to. Though still physically present at mass, Maria felt emotionally detached, going through the motions without belief. They both wondered if they would ever find peace again. However, within their respective religious communities, they found unexpected support. Edgar confided in an old mentor, a pastor who had known him since childhood. Instead of condemnation, the pastor listened. He shared his own struggles with doubt, emphasizing that faith was not the absence of questions but the willingness to seek answers. The conversations did not provide Edgar with immediate certainty, but they gave him space to wrestle with his beliefs without fear of judgment.

Maria found solace in a support group within her church, which gathered individuals who had also faced grief and doubt. There, she met others who had lost loved ones and had struggled to reconcile their pain with their faith. She learned that questioning did not mean abandoning and slowly embraced the idea that faith could coexist with uncertainty.

Both Edgar and Maria took different paths toward spiritual peace. James began exploring theology, studying different religious philosophies, and engaging in deep discussions. He came to accept that faith was not about having all the answers but about embracing the mystery. Maria, on the other hand, found comfort in volunteering at a hospice center, supporting those in grief, and, in doing so, rediscovering compassion at the heart of her faith. Their journeys were not identical, nor were they linear. There were moments of doubt, setbacks, and frustrations. However, in the end, both found a way to reconcile their struggles and rediscover a sense of spiritual peace, not by erasing their doubts but by learning to live with them. Their stories serve as reminders that faith is not a destination but a journey that is most meaningful when shared with others who understand the weight of uncertainty and the power of hope.

In a study done in 2021 by Springtide Institute (Packard, 2021), young people were asked to participate in a survey about particular crossroads that could evoke uncertainty. Some of the questions asked were: “making decisions about the future; sudden events like death or disease; relationship transitions like breakups, divorce, or new partnerships; major events like graduating, moving, or starting a career.” Additionally, they were asked about the degree to which they felt “stressed, anxious, worried, confused, or scared when navigating uncertain circumstances?” Conversely, they were also asked "whether they felt calm, confident, excited, or optimistic about whatever was coming next?" The results of the survey indicated that young people are experiencing uncertainty. Forty percent said they had experienced trauma, sixty-two percent said they had experienced emotional distress as a result of a challenging event, and twenty-eight percent reported experiencing a challenging event that was causing them to feel uncertain, making them feel unsettled, uncomfortable, or stressed at the time they were completing the survey. Twenty-five percent of the respondents who identified as male stated that they were experiencing uncertainty at the time of the survey, 31% of female respondents, and 32% of non-binary respondents, respectively. The data indicated that" there is a disconnect between young people and religious institutions. Meanwhile, the disconnect is not simultaneous with a loss of interest in spiritual and religious questions among young people or even a loss of faith.

The 2021 Springtide Research Institute study, *The State of Religion & Young People: Navigating Uncertainty*, revealed that while nearly 75% of young people identify as religious or spiritual, they often do not turn to faith communities during moments of stress or uncertainty. Instead, they are exploring new, individualized ways of constructing meaning and encountering the divine, a concept the study calls "Faith Unbundled." This approach reflects a shift away from traditional religious structures, with young people drawing on diverse sources to shape their spiritual lives.

The study highlights a significant gap in support, as only 10% of young people reported being contacted by a faith leader in the past year. It underscores the importance of rethinking how faith communities engage with younger generations, offering insights and frameworks to bridge this disconnect. Ultimately, the findings call for a more inclusive and adaptive approach to supporting young people as they navigate life's uncertainties.

Exploring the Experience of a Crisis of Faith: A Qualitative Study" by Claire D. Webb

Webb (2019) conducted a qualitative study on the occurrences of a crisis of faith. This qualitative study examines the personal experiences of individuals undergoing a crisis of faith within the Christian tradition. Using grounded theory methodology, Webb found that crises are closely linked to one's relationship with God, fluctuating based on self-acceptance and acceptance of God's love. The study suggests that experiencing a crisis of faith can lead to a deeper relationship with God and spiritual growth, supporting the concept of God as an attachment figure with whom individuals interact differently depending on their attachment styles and past experiences.

The Journey from Tragedy to Hope: The Experience of Christian Undergraduates" by David M. Johnstone

In another case study, Johnstone (2014) studied *The Journey from Tragedy to Hope: The Experience of Christian Undergraduates*. His case study focuses on seven student leaders at an evangelical university who had experienced tragic or traumatic events in their lives. Despite these challenges, they maintained a hopeful outlook. The research identified common themes in their narratives. The themes expressed included the importance of relationships, community, and individual belief. Notably, the study highlights the significance of having someone assist in deriving meaning or purpose from their experiences. It suggests that educators can be crucial in supporting students' processing of traumatic events.

In a study titled *The Existential Quest: Doubt, Openness, and the Exploration of Religious Uncertainty,* Arrowood, Vail, and Cox report (2022), this study investigates how individuals with a "quest" orientation who view religion as a journey involving questioning and exploration manage existential concerns. Through a series of six studies, the researchers found that such individuals experience heightened death-related anxiety and cognitive vulnerability when confronted with mortality reminders. However, when encouraged to explore their doubts actively, these individuals exhibited increased cultural openness and reduced reliance on supernatural beliefs. This suggests that engaging with religious uncertainty can lead to personal growth and broader cultural understanding.

"Navigating Faith Transitions: A 4-Year Longitudinal Examination of Religious Deidentification Among LGBTQ+ Latter-day Saints"

In a Longitudinal Examination of Religious Deidentification Among LGBTQ+ **Latter-day Saints" (2025),** the study sought to understand the engagement of participants who identify with the LGBTQ+ community. The longitudinal study followed 164 LGBTQ+ individuals who were active members of the Church of Jesus Christ of Latter-day Saints over four years. The research revealed that approximately two-thirds of participants reduced their religious affiliation, with 40% fully disaffiliating from the church. Younger participants tended to de-identify more rapidly, and increased engagement with LGBTQ+ communities was a significant predictor of decreased religious service attendance. Notably, the study found no adverse effects on mental health or well-being associated with religious deidentification, indicating that this process can be a normative and healthy aspect of identity development for LGBTQ+ individuals(.PubMed).

These studies underscore the complex interplay between personal identity, community engagement, and the process of navigating religious uncertainty.

Understanding the engagement of young adults and diverse groups, including the LGBTQ+ community, in faith commitment and religious communities is crucial for fostering inclusivity, addressing spiritual needs, and promoting meaningful participation. Religious affiliation and spiritual identity are significant in personal development, social belonging, and ethical reasoning. Nevertheless, young adults and LGBTQ+ individuals often encounter unique challenges that impact their religious engagement. Many young adults experience transitions in their belief systems due to exposure to diverse ideologies and shifting cultural norms. At the same time, LGBTQ+ individuals may struggle with tensions between their identities and traditional religious teachings (Bowman et al., 2021). Research indicates that inclusive religious spaces that embrace diversity enhance well-being and provide crucial support for marginalized groups, allowing them to engage authentically with faith communities (Rodriguez & Ouellette, 2019). Scholars and religious leaders can develop approaches that accommodate evolving spiritual needs and create more welcoming, equitable environments for all individuals by examining how these groups interact with faith institutions.

Conclusion

Religious uncertainty, as seen through the lives of Edgar, Maria, and countless others, is not a sign of spiritual failure. it is often the birthplace of deeper, more authentic faith. The journey through doubt, grief, or identity conflict doesn't always lead back to familiar doctrines or institutions. Sometimes it carves out new paths of understanding, where mystery is embraced, traditions are reimagined, and spirituality becomes more personally meaningful. The case studies and research examined here reveal a powerful truth: questioning does not weaken faith; it refines it. Whether it's a college student wrestling with inherited beliefs, a grieving sibling searching for answers, or an LGBTQ+ believer navigating spiritual belonging, each story reflects the evolving nature of how people make meaning in today's world.

This chapter affirms that religious uncertainty is not something to be feared or silenced. Instead, it invites communities, educators, and leaders to engage with it honestly and compassionately. In a time when more individuals, especially young adults are unbundling faith from rigid systems, there is both challenge and opportunity. Supporting this quest for meaning with openness and empathy may not return everyone to the same pews, but it can foster spiritual integrity, emotional healing, and a renewed connection with the divine—one shaped not by certainty, but by courage, curiosity, and community.

Chapter 13

BUILDING RESILIENT EDUCATIONAL SYSTEMS

Building resilient education systems is essential for a nation's long-term development and stability. Education serves as the cornerstone for economic growth, social cohesion, and individual empowerment. Resilient systems ensure continuity in learning despite disruptions such as natural disasters, pandemics, or political instability. They also address inequalities, providing equitable access to quality education for all citizens, regardless of socio-economic status or geographic location. By fostering adaptability, innovation, and inclusivity, resilient education systems equip nations to tackle emerging challenges and opportunities in a rapidly changing world.

Education is a fundamental human right and a cornerstone for societal development. However, despite efforts to provide inclusive education for all, disparities persist due to socioeconomic status, geographic location, and systemic inequalities. The need for equitable and resilient education systems has never been more urgent, especially due to global challenges such as the COVID-19 pandemic. This chapter explores strategies for fostering equity and resilience in education, the role of technology and innovation, and the future trends and challenges that lie ahead.

Creating fair and resilient education systems requires comprehensive policy reforms and inclusive curriculum development. Governments must implement policies that target funding toward underprivileged schools and marginalized communities, addressing disparities in education. Developing curricula that reflect diverse cultural perspectives can foster a sense of belonging and engagement among students from various backgrounds. Educators play a critical role in these efforts, and equipping them with skills to manage diverse classrooms and adapt to different learning needs is essential. Ongoing professional development programs focusing on modern teaching techniques, cultural competency, and the effective use of technology can empower teachers to meet the challenges involving education systems. Infrastructure investment is another vital component, as many underserved regions lack basic educational facilities and resources. Creating well-equipped classrooms, libraries, and sanitation facilities can significantly improve learning conditions for students in these areas.

Community and Parental Engagement

Community and parental engagement refers to the active involvement of both parents and the broader community in supporting and enhancing the educational experience of students. This includes activities like open communication with educators, participation in school initiatives, collaboration in decision-making, and creating a nurturing and inclusive environment. Such engagement is vital for building resilient schools because it fosters strong partnerships that help schools adapt to challenges, address diverse student needs, and ensure continuity in learning during disruptions. By encouraging shared responsibility and leveraging collective resources, schools can create a supportive network that promotes the academic, social, and emotional well-being of students while fostering a sense of belonging and trust within the school community.

Community and parental engagement are equally important in creating supportive learning environments. Schools should build strong relationships with parents and communities, while community-led initiatives such as tutoring, mentorship, and supplementary learning resources can help bridge educational gaps. Flexible learning models, such as blended approaches that combine in-person and online instruction, can accommodate diverse needs and circumstances. Alternative pathways, such as vocational training and competency-based education, offer viable career options for students and address the varied demands of the modern workforce.

The Use of Technology for Transformation

The use of technology is an urgent necessity to drive school transformation due to its ability to modernize educational practices, enhance learning outcomes, and prepare students for a rapidly evolving digital world. Technology facilitates personalized learning, expands access to educational resources, and fosters collaboration among students and educators. It also equips schools to respond effectively to disruptions, such as pandemics or natural disasters, by enabling remote learning and virtual classrooms. Furthermore, integrating technology into education systems nurtures digital literacy, which is critical for students to thrive in future academic, professional, and social settings. As schools aim to foster adaptability and innovation, leveraging technology becomes a pivotal step in creating resilient and future-ready institutions.

Technology is a transformative tool that can bridge gaps in access, improve educational quality, and ensure continuity in the face of disruptions. Integrating digital platforms like online classrooms and learning management systems allows students and educators to connect and collaborate remotely. This became especially evident during the COVID-19 pandemic, where virtual learning became a lifeline for education. The use of Artificial Intelligence and machine learning can personalize learning experiences, adapting lessons to individual students' needs and progress. Additionally, mobile apps, e-books, and digital libraries provide resources that are accessible anytime and anywhere, removing barriers for students in remote or underserved areas. Investing in internet infrastructure and providing devices to students ensures inclusivity in technological advancements. Technology and innovation are transformative forces in education, providing new ways to enhance learning experiences and reduce disparities. Digital learning platforms, including online courses and open educational resources (OER), offer access to quality education regardless of location and alleviate financial barriers for students. Artificial intelligence enables personalized learning experiences through intelligent tutoring systems that provide real-time feedback, helping students grasp complex concepts. Virtual and augmented reality further enhances engagement and comprehension, offering immersive learning experiences in technical fields like science and medicine. Assistive technologies, such as screen readers and speech-to-text tools, support students with disabilities, while language translation applications aid multilingual learners. Data analytics provides insights for identifying learning gaps and designing targeted interventions, ensuring a more effective and inclusive educational approach.

Looking ahead, several trends and challenges will shape the future of education. The rapid pace of technological advancement demands a stronger emphasis on lifelong learning, with education systems adopting models that include micro-credentials and upskilling opportunities. Global collaboration and knowledge-sharing initiatives can enhance cross-border efforts to address educational disparities and share best practices. However, achieving equity in digital access remains a critical challenge, as many students still lack reliable internet and devices. Investments in digital infrastructure are essential to bridge this divide. Ethical considerations in the use of AI and data must also be addressed, with regulations ensuring privacy and unbiased application. Finally, resilience against future disruptions, as highlighted by the COVID-19 pandemic, requires education systems to be adaptable and prepared for emergencies. Scalable online learning solutions and robust contingency plans will be key to future-proofing education systems.

By integrating policy reforms, technological innovation, and community engagement, education can become more inclusive and adaptive to society's evolving needs. Addressing systemic barriers, ensuring digital equity, and prioritizing lifelong learning will be pivotal in creating a sustainable and fair learning environment for all. Through these efforts, education can empower individuals worldwide and contribute to a more equitable future.

Building resilient educational systems is a cornerstone for fostering sustainable development and equitable progress in societies worldwide. Resilience in education ensures that systems can adapt to challenges such as pandemics, natural disasters, political instability, and economic downturns without compromising quality or accessibility. It begins with addressing the foundational needs of infrastructure, teacher training, and curriculum design that are both flexible and inclusive. Integrating technology, promoting digital literacy, and ensuring access to online learning resources enable continuity in education, regardless of unforeseen disruptions.

Mental Health in Resilient Educational Systems

Resilient mental health initiatives are crucial within resilient educational systems because they support the emotional and psychological well-being of students, educators, and the broader school community. Mental health is foundational to learning and personal development; when students and staff feel secure, valued, and supported, they are better equipped to thrive academically and socially. Such initiatives help schools address challenges like trauma, anxiety, and stress, particularly during times of crisis or disruption. By creating a framework that prioritizes mental health, schools can foster an inclusive and empathetic environment that enhances resilience, improves learning outcomes, and builds a culture of mutual support and understanding.

Mental health is the basis upon which academic success and overall well-being are built, yet it is often overlooked in educational frameworks. To build resilient systems, mental health support must be integrated into schools at every level. This includes training teachers and staff to recognize signs of stress, anxiety, and other mental health challenges in students. Counseling services and mental health professionals should be readily available, offering both one-on-one support and group workshops. Promoting a positive school culture that emphasizes inclusivity, kindness, and open communication can create a safe environment where students feel valued and heard. Additionally, mental health education should be part of the curriculum, teaching students about emotional resilience, coping strategies, and how to seek help when needed. By addressing mental health proactively, educational systems can reduce stigma, improve retention rates, and help students navigate challenges both inside and outside the classroom. Remember, mental health support is just as vital for educators, whose well-being directly impacts their ability to nurture and inspire their students. Further, prioritizing mental health support for students and educators fosters a nurturing environment where everyone can thrive, even in adversity. Partnerships between governments, private sectors, and local communities can pool resources and expertise to create robust systems that are well-equipped to handle crises. Policies that emphasize equity ensure that marginalized groups—such as those in rural areas, low-income families, or with disabilities—are not left behind. Incorporating localized approaches tailored to specific cultural, geographical, and social contexts amplifies the effectiveness of these systems.

Moreover, embedding a culture of lifelong learning encourages individuals to adapt and grow in a rapidly changing world, contributing to overall societal resilience. Continuous investment in research and data-driven decision-making helps identify gaps and opportunities for improvement, ensuring that educational systems evolve proactively. Building resilience also includes preparing students with critical thinking, problem-solving, and collaboration skills that empower them to navigate uncertainty in their futures. By prioritizing sustainability, we can ensure educational frameworks that not only endure but also excel over time. Ultimately, resilient educational systems serve as a backbone for thriving communities, innovation, and a more just and equitable world, proving that education is both a right and a catalyst for transformative change.

Student mental health is critical to their learning and personal development. As Burns and Rapee (2019) emphasize, "a mentally healthy school environment enhances students' capacity to learn, solve problems, and connect socially with their peers." Similarly, Weare (2015) notes that "addressing mental health within schools is not an added burden but rather an essential part of fostering effective learning and development." [3]

In today's rapidly changing and often unpredictable world, developing resilient schools with comprehensive mental health programs integrated into their very fabric is not just important—it is urgently essential. Schools are more than places of academic instruction; they are critical environments where children grow, develop, and prepare for life's challenges. Mental health initiatives embedded into school culture not only support students in

---

[3]Burns, J., & Rapee, R. M. (2019). *Mental health in schools: Engaging teachers and students in a resilience-based model.* Springer.
Weare, K. (2015). *Promoting mental, emotional, and social health: A whole-school approach.* Routledge.

overcoming barriers to learning but also foster a compassionate, inclusive atmosphere where educators, families, and communities thrive together. As Weare (2015)[4] observes, "a whole-school approach to mental health not only boosts academic achievement but also promotes emotional resilience essential for lifelong success." [5]Burns and Rapee (2019) similarly highlight that "mental health initiatives in schools empower students and staff to navigate crises, trauma, and everyday stress with adaptability and strength." By establishing a foundation of mental health support that permeates every aspect of the school experience, educational institutions can transform into havens of stability, growth, and empowerment, equipping future generations to thrive in even the most turbulent times.

Conclusion

Building resilient educational systems is essential for ensuring continuity, equity, and quality in a world of constant change. This chapter highlighted the importance of inclusive policies, strong infrastructure, teacher training, and culturally responsive curricula. Community and parental engagement strengthen schools' ability to adapt, while flexible learning paths and vocational options make education more accessible and relevant.

Technology plays a critical role in expanding access and personalizing learning, but only if digital equity is addressed. Equally important is mental health—resilient systems must support the emotional well-being of both students and educators. By integrating these elements, education becomes not only a tool for academic growth but a foundation for societal resilience and long-term progress.

---

[4] Burns, J., & Rapee, R. M. (2019). *Mental health in schools: Engaging teachers and students in a resilience-based model*. Springer.

[5] Weare, K. (2015). *Promoting mental, emotional, and social health: A whole-school approach*. Routledge.

Chapter 14

SHAPING HOPEFUL POLITICAL FUTURES

The importance of civic engagement and participation, along with thoughtful policy recommendations to nurture hope and reduce fear, cannot be overstated. Throughout history, the pursuit of a just and hopeful political landscape has been central to social movements, scholars, and leaders alike. This shared vision prioritizes equality, justice, and the common good, empowering citizens through active involvement to help shape a better future. Yet achieving this vision requires more than ideals—it demands the full and active participation of individuals at every level of governance. Civic engagement stands as a cornerstone of democracy, playing an essential role in creating a political environment where fairness and optimism can flourish. This chapter explores the power of civic involvement, the societal forces that either feed fear or foster hope, and policy actions that can pave the way toward a more just and inclusive society.

Civic engagement refers to the participation of individuals in the political process—whether through voting, activism, community organizing, or other forms of involvement. It is the lifeblood of democracy and a key mechanism through which ordinary people influence policies that affect their daily lives. History has shown that even the most marginalized voices gain power through collective political action. The Civil Rights Movement in the United States, for example, led by figures such as Martin Luther King Jr., demonstrated how civic participation could drive systemic change and uphold justice and fairness.

Importantly, civic engagement is not a one-time event—it is an ongoing process. It builds a sense of ownership and agency over the direction of political decisions. A healthy democracy depends on an informed and active electorate—people who hold leaders accountable and advocate for policies rooted in equality, inclusion, and justice. Without consistent citizen participation, democratic systems risk drifting toward authoritarianism, corruption, and widespread public disillusionment. Therefore, fostering civic involvement is not just idealistic; it is essential to maintaining a system that meets the needs of all.

A just political landscape is one where every individual—regardless of race, gender, economic status, or background—has equal access to opportunity and is treated with dignity. Justice goes beyond legal fairness; it encompasses the social and economic conditions that enable people to thrive. Creating such a society involves intentionally dismantling systemic inequities that have long disadvantaged marginalized groups, including communities of color, women, and low-income families.

To make this vision a reality, structural reforms must address the root causes of injustice. Tackling issues like income inequality, healthcare access, education disparity, and institutional racism is not optional—it is necessary. This includes adopting progressive taxation, ensuring universal healthcare, and reforming public education to make it more equitable. Equally important is fostering inclusive public discourse, where historically silenced voices are heard and valued. Justice is not just about legal systems—it is about inclusive systems that reflect the full spectrum of society.

Fear, however, can derail these efforts. In many parts of the world, fear has become a potent force in shaping political decisions. Whether tied to economic instability, crime, or perceived cultural threats, fear is often exploited by political actors who use it to divide populations and foster hostility. This fear-driven environment discourages civic participation and makes room for extremist ideologies that offer simplistic solutions to complex societal problems.

Yet fear need not dominate political life. When hope is nurtured and a collective sense of purpose is fostered, societies can push back against the paralyzing effects of fear. Hope inspires action. It brings people together and builds a belief that progress is possible. When communities organize around shared values and goals, they reclaim their power from fear and contribute meaningfully to political life. In this way, hope becomes the antidote to apathy and disengagement.

To truly cultivate a more hopeful and just political culture, specific policies must be implemented to encourage participation and reduce fear. These policies should be comprehensive and inclusive, creating tangible improvements in citizens' lives that inspire trust in public institutions and renewed confidence in democratic processes.

First, enhancing civic education is critical. A well-informed public is the foundation of a functioning democracy. Civics should be a mandatory part of every school curriculum—not just teaching the structure of government but encouraging critical thinking, community involvement, and an understanding of one's rights and responsibilities. Accessible educational platforms for adults should also be prioritized, ensuring that people of all ages can engage meaningfully in civic life.

Second, expanding access to voting is fundamental. Practical steps such as same-day registration, early voting options, mail-in ballots, and the removal of restrictive voter ID laws can significantly increase participation—especially among historically disenfranchised groups. Ensuring that every eligible citizen can vote without undue burden is a key step in revitalizing democratic engagement.

Third, creating forums for deliberative democracy can bring people into meaningful dialogue. Citizens' assemblies, town hall meetings, and community dialogues give individuals a direct voice in policymaking. These participatory platforms foster transparency, trust, and collaboration—helping citizens feel that their concerns are not only heard but acted upon.

Fourth, economic and social equity must be addressed to reduce the root causes of fear. Economic insecurity feeds anxiety and division, making individuals more susceptible to fear-based narratives. Governments can reduce this fear by ensuring equitable access to healthcare, education, housing, and income support. When citizens feel secure in their basic needs, they are more likely to invest time and energy into public life.

Lastly, inclusive political messaging plays a powerful role in shaping the civic atmosphere. Leaders must choose language that unites rather than divides. When public figures highlight common ground and shared aspirations, they create a political climate that welcomes participation and cooperation. Messaging rooted in empathy and collective progress can help heal divisions and remind people of their shared humanity.

The pursuit of a hopeful and just political society is not abstract or unattainable. It is a collective endeavor that can be realized through intentional civic engagement, meaningful policy reform, and a commitment to reducing fear. Every citizen has a role to play in shaping the future. Civic engagement is not merely a right—it is a vital responsibility. Through education, access, equity, and unity, we can build a society in which hope overcomes fear and justice becomes a reality for all.

Conclusion

The pursuit of a hopeful and just political society is not abstract or unattainable. It is a collective endeavor that can be realized through intentional civic engagement, meaningful policy reform, and a commitment to reducing fear. Every citizen has a role to play in shaping the future. Civic engagement is not merely a right—it is a vital responsibility. Through education, access, equity, and unity, we can build a society in which hope overcomes fear and justice becomes a reality for all.

But this work does not happen overnight. It requires persistence, courage, and a willingness to believe that our shared actions, no matter how small, can shift the course of history. It means speaking out even when it's inconvenient, showing up even when it's uncomfortable, and staying engaged even when progress feels slow. It means rejecting apathy and resisting fear, not just for ourselves, but for future generations who will inherit the systems we choose to build or neglect.

The road to a fairer, more compassionate political landscape is long, but it is walkable, step by step, voice by voice, vote by vote. And as we walk it together, guided by justice and fueled by hope, we carry the power not only to imagine a better world, but to create one.

Chapter 15

FOSTERING INTERFAITH UNDERSTANDING AND COOPERATION

In the world today, where societies are more interconnected than ever before, the importance of interfaith dialogue cannot be overstated. As the global community faces pressing challenges such as political instability, environmental crises, and social inequalities, fostering understanding between different religious groups is vital to creating a more harmonious world. Interfaith dialogue—the open, respectful, and meaningful exchange between people of different religious traditions—has the potential to build bridges, promote peace, and encourage shared values. This essay explores various initiatives and movements that promote interfaith dialogue, highlights success stories of interfaith collaboration, and discusses the future of religion in an increasingly diverse and interconnected world.

Initiatives and Movements Promoting Interfaith Dialogue

The need for interfaith dialogue has grown significantly as religious diversity has increased across the world. Several initiatives and movements have arisen to foster conversations that transcend religious boundaries, aiming to promote mutual understanding, respect, and collaboration.

1. **The Parliament of the World's Religions: the Parliament of the World's Religions** (PoWR) is one of the largest and most well-established global platforms for interfaith dialogue. Since its inception in 1893, PoWR has held biennial gatherings to bring together religious leaders, scholars, and activists to discuss issues of common concern, such as global peace, poverty, human rights, and climate change. Parliament provides a space for exchanging ideas and encourages religious communities to work together to address the world's pressing issues. Its mission is to foster a world in which

people of different faiths can coexist peacefully and contribute collectively to global well-being.

2. **The United Religions Initiative (URI) The United Religions Initiative**, founded in 2000, is a global grassroots movement that promotes interfaith cooperation. URI's network spans over one hundred countries, with over 1,000 member organizations. The initiative seeks to promote religious tolerance and create alliances across different faiths to address social and environmental justice issues. URI's approach focuses on creating locally based action groups, known as Cooperation Circles, which work to bring people of various faiths together for dialogue, advocacy, and collaborative action on global issues like poverty, environmental sustainability, and human rights.

3. **The Interfaith Youth Core (IFYC) The Interfaith Youth Core**, based in the United States, is a prominent initiative that focuses on engaging young people in interfaith dialogue. Founded in 2002 by Eboo Patel, the IFYC works to inspire young people to work together across religious divides to address shared challenges. The organization provides training, resources, and leadership opportunities for young people to become catalysts for interfaith cooperation in their communities. Through programs like the Interfaith Leadership Institutes and the Interfaith Leadership Council, the IFYC empowers youth to create inclusive communities and forge connections across faith traditions.

4. Religions for Peace Religions for Peace (RfP), founded in 1970, is another key initiative aimed at fostering interfaith dialogue and collaboration. It is the world's largest coalition of religious representatives working together to promote peace. RfP brings together religious leaders and communities to address conflict, human rights abuses, and social injustices, as well as to promote sustainable development. Its efforts include organizing interfaith peace-building missions, dialogues, and collaborative projects aimed at fostering understanding and reducing religiously motivated violence.

Success Stories of Interfaith Collaboration While interfaith dialogue remains an ongoing process, several success stories highlight the positive outcomes of religious cooperation in various regions and sectors. These stories exemplify what is possible when people of different faiths unite for the common good.

1. **The Abrahamic Reunion in the Middle East** The Abrahamic Reunion, a network of Jewish, Christian, and Muslim religious leaders in the Middle East, is a remarkable example of interfaith collaboration. Founded in 2007, the Abrahamic Reunion aims to promote peace and understanding in the conflict-ridden region. The group of religious leaders has been instrumental in organizing meetings and dialogues between communities that are often at odds, advocating for peace, and working to overcome the deep divisions caused by centuries of conflict. This initiative has proven to be a significant step toward healing religious tensions in the region and building long-term peace.

2. **The Interfaith Cooperation Forum in Asia:** The Interfaith Cooperation Forum (ICF), established in 2003, is a network of over fifty organizations from various Asian countries, including India, Pakistan, Nepal, and Sri Lanka. The forum addresses social, political, and religious conflict in the region by encouraging interfaith dialogue and collaboration. ICF's efforts in promoting peace and social justice have led to successful peace-building projects, particularly in conflict zones. By collaborating with grassroots communities, ICF has fostered collaboration between Hindus, Muslims, Buddhists, and Christians, contributing to social harmony in a region often marked by sectarian strife.

3. **The U.S. Interreligious Council's Advocacy for Religious Freedom:** In the United States, the U.S. Interreligious Council has played a significant role in advocating for religious freedom and interfaith understanding. One of the Council's notable achievements was its involvement in the passage of the International Religious Freedom Act of 1998, which aimed to promote religious liberty globally. The Council's work continues to foster interfaith cooperation on domestic and international levels, with a focus on religious tolerance, human rights, and social justice.

4. The Interfaith Friendship Network in South Africa: South Africa, with its history of apartheid and religious diversity, has witnessed several interfaith initiatives aimed at fostering reconciliation and social cohesion. The Interfaith Friendship

Network, founded in the post-apartheid era, is one such initiative. It brings together people of various faiths, including Christianity, Islam, Judaism, and traditional African religions, to work toward national healing and social justice. This network's efforts have contributed to the healing process in a society scarred by division and violence, demonstrating the power of interfaith dialogue in promoting national unity.

The Future of Religion in a Diverse and Interconnected World: as the world becomes more interconnected, the role of religion in shaping societies and global politics will continue to evolve. The future of religion in a diverse world will be shaped by both the challenges and opportunities that arise from increasing interfaith interactions.

1. The Rise of Global Interfaith Networks In the future, the growing interconnectedness of societies will lead to expanding global interfaith networks. These networks will provide platforms for religious communities to engage with one another on issues such as climate change, social justice, and global peace. Interfaith organizations will become even more essential in creating spaces for respectful dialogue and collaboration, helping to address global challenges through collective action.

2. Technology and Interfaith Dialogue With the rise of digital communication and social media, technology will play a crucial role in the future of interfaith dialogue. The ability to connect with people from distinct parts of the world instantly allows for more inclusive conversations. Online platforms can serve as a space for religious leaders, scholars, and individuals to engage in dialogue, exchange ideas, and challenge stereotypes. However, these platforms also present challenges, as they can sometimes amplify intolerance and hatred. The future will require intentional efforts to harness technology to promote positive interfaith exchanges while combating religious extremism and intolerance online.

3. Religious Pluralism as a Catalyst for Social Cohesion

Religious pluralism—the coexistence of multiple religious traditions in a single society—will continue to be a driving force in promoting social cohesion. In countries with diverse religious populations, fostering understanding and respect between religious groups will be essential to maintaining social stability. Education will play a vital role in this process, as schools and universities can serve as key spaces for young people to engage in interfaith dialogue and learn about different religious traditions. By emphasizing common values such as compassion, justice, and peace, future generations can be equipped to live harmoniously in a diverse and interconnected world.

2. **Challenges of Secularism and Religious Fundamentalism:** As religious pluralism grows, so too do the challenges posed by secularism and religious fundamentalism. In many parts of the world, secularization is leading to a decline in religious adherence, while religious fundamentalism is on the rise in others. These opposing forces create tension between individuals and communities who seek either to diminish the role of religion in public life or to assert the supremacy of one religion over others. The future will require nuanced approaches to balance religious freedom and secular values while ensuring that religious communities continue to contribute positively to society.

Interfaith dialogue fosters understanding and peace in a diverse and interconnected world. Initiatives such as the Parliament of the World's Religions, the United Religions Initiative, and the Interfaith Youth Core promote cooperation between different faith communities. Success stories like the Abrahamic Reunion and the Interfaith Cooperation Forum highlight the positive impact of collaboration in conflict-prone regions. As the world continues to grow more interconnected, the future of religion will be shaped by both the challenges and opportunities presented by increasing religious pluralism and the need for continued interfaith dialogue. With the right approach, religion can serve as a unifying force, contributing to global peace and social harmony.

CONCLUSION

The Interconnectedness of Hope, Fear, and Uncertainty in Education, Politics, and Religion

Hope, fear, and uncertainty are emotional forces that shape human experience in profound ways. These emotions interact with each other, creating a complex web of influences that shape individual and societal actions. In the contexts of education, politics, and religion, these three emotions are interwoven, influencing how individuals perceive and engage with the world around them. In education, the desire for knowledge and personal growth is often accompanied by fear of failure or uncertainty about the future. In politics, hope for progress and change can coexist with fear of societal collapse or the unknown. Religion, with its promise of hope for eternal life or divine intervention, is often accompanied by the fear of judgment or uncertainty about one's spiritual fate. This essay explores how hope, fear, and uncertainty are interconnected within these three domains and how their interplay shapes our attitudes, decisions, and actions.

Hope, Fear, and Uncertainty in Education

Education is often seen as a pathway to personal and societal progress. It is a domain where hope plays a significant role: students hope to learn, grow, and improve their lives. For educators, there is the hope of shaping young minds and preparing students for the future. However, this hope is frequently accompanied by fear and uncertainty. Students may fear failure, judgment, or not living up to expectations, whether from themselves, their peers, or their teachers. This fear can create anxiety and hinder learning, even as hope for success drives them forward.

The uncertainty that accompanies education is also significant. Both students and educators face the unknowns of the future. For students, this uncertainty often centers around their future careers, societal roles, and the implications of their educational achievements. Will their chosen path lead to success or failure? For educators, the uncertainty stems from the constantly shifting landscape of education policy, student needs, and technological advancements. These factors contribute to the emotional and intellectual tension in educational settings.

The interplay of hope, fear, and uncertainty in education is not inherently negative. In fact, it can drive individuals to push themselves harder, seek innovative solutions, and innovate. For example, a student may fear failure but simultaneously hope that their hard work will lead to success. Similarly, educators may feel uncertain about how to reach every student best, but their hope for positive change keeps them committed to their work. The challenge lies in managing these emotions to create an environment where learning can flourish rather than being stifled by excessive fear or uncertainty.

Hope, Fear, and Uncertainty in Politics

In the political arena, hope, fear, and uncertainty play crucial roles in shaping public opinion and guiding actions. Politics, by nature, involves competing visions of the future, and citizens and leaders alike are motivated by hopes for a better tomorrow. Political candidates often appeal to voters' hopes, promising a brighter future, whether through economic reform, social justice, or improved healthcare. These hopeful messages are central to political campaigns, as they inspire action, mobilize supporters, and provide a vision of what the future could look like.

However, politics is also fraught with fear and uncertainty. Fear of the unknown, fear of losing rights, and fear of instability can drive political decision-making. In times of economic or social crises, fear often becomes a dominant force in political discourse. Citizens may fear that the political establishment is not capable of addressing the challenges at hand, or they may fear that their values and beliefs are under threat from the opposition. This fear can lead to polarization, as individuals gravitate toward leaders who promise to safeguard their interests and alleviate their fears.

Uncertainty in politics is particularly pronounced in times of transition or upheaval. Elections, policy changes, and global events introduce uncertainty about the future. Voters may feel unsure about whether the promises made by political leaders will be fulfilled or whether they will have a lasting impact. Similarly, political leaders must navigate the uncertainty of shifting public opinion, international relations, and unforeseen crises. In such environments, hope for progress and change must contend with the fear of setbacks and the uncertainty of what lies ahead.

The interconnectedness of hope, fear, and uncertainty in politics creates a dynamic that can lead to both positive and negative outcomes. When these emotions are harnessed effectively, they can drive progress and encourage collective action. However, when fear dominates, it can lead to divisiveness and regression. The challenge in politics is to balance these emotions in a way that fosters cooperation, inclusivity, and long-term solutions rather than relying solely on fear-based rhetoric or ungrounded optimism.

Hope, Fear, and Uncertainty in Religion

Religion, like education and politics, is a domain where hope, fear, and uncertainty are deeply interwoven. In many religious traditions, hope is central. Hope for salvation, eternal life, or divine intervention provides comfort and guidance to individuals facing life's challenges. Faith in a higher power often offers believers a sense of hope that transcends the material world and provides meaning in times of suffering.

Yet, alongside this hope, fear and uncertainty are ever-present. In many religious contexts, fear of divine judgment or fear of sin can create a sense of unease. Believers may fear that they will fall short of the standards set by their faith, leading to feelings of guilt, shame, or anxiety. The fear of eternal punishment or separation from the divine is a powerful motivator for many religious adherents, pushing them to live according to specific teachings and practices.

Uncertainty also plays a significant role in religion. While hope is often rooted in faith, there remains an inherent uncertainty about the afterlife, the existence of a higher power, and the fate of the soul. Even the most devout believers may experience moments of doubt or confusion, questioning the teachings of their faith or the meaning of their experiences. This uncertainty can lead to spiritual growth as individuals wrestle with existential questions, but it can also create anxiety and doubt, particularly during challenging times.

The interconnectedness of hope, fear, and uncertainty in religion is both a source of strength and a potential source of conflict. The hope for salvation or enlightenment is a powerful force, but it can be tempered by the fear of not measuring up to religious ideals or the uncertainty of what lies beyond this life. As in education and politics, the key is to manage these emotions in a way that encourages growth, understanding, and compassion rather than allowing them to breed anxiety, exclusion, or division.

Call to Action

The interconnectedness of hope, fear, and uncertainty in education, politics, and religion is a testament to the complexity of human experience. These emotions shape our perceptions, decisions, and actions, influencing both individual lives and the broader societal context. While hope drives us to pursue knowledge, change, and spiritual fulfillment, fear and uncertainty can hold us back or push us toward division and regression. Understanding the interplay of these emotions in various spheres of life is essential for fostering environments that promote growth, cooperation, and compassion.

In education, we must strive to create learning environments that support students' hopes while alleviating their fears and uncertainties. Educators and policymakers must acknowledge the emotional challenges students face and provide the tools and resources necessary to navigate those challenges. In politics, leaders must recognize the power of hope to inspire and unite while also addressing the fears and uncertainties that divide us. Rather than relying on fear-based rhetoric, political leaders should cultivate hope and work toward solutions that reduce uncertainty and promote collective well-being. In religion, faith leaders should offer guidance that nurtures hope and spiritual growth while acknowledging the fear and uncertainty that come with the human condition. By embracing these emotions and understanding their interconnectedness, we can create a more compassionate, resilient, and unified society.

The call to action is clear: we must work together to foster a world where hope can flourish without being overshadowed by fear and uncertainty. In education, politics, and religion, we could guide ourselves and future generations toward a more hopeful, compassionate, and understanding future. Let us embrace the complexities of human experience, using them as a foundation for growth, unity, and positive change.

EPILOGUE

As the final words of this book are written, the nation stands at a crossroads—one defined not just by the division we have explored but by the possibilities that lie ahead. The journey through our nation's concerns has been both sobering and enlightening. We have dissected the challenges that pull us apart, from economic inequality and racial tensions to political polarization and social unrest. We have explored the fears that grip communities, the hopes that drive movements, and the possibilities for change that rest in the hands of everyday people. Now, the question remains: where do we go from here?

Rising above the challenges of our time requires more than rhetoric, more than slogans, and certainly more than performative gestures. It requires action, courage, and a willingness to engage with those who think differently than we do. We must ask ourselves: Are we willing to listen? Are we willing to compromise for the greater good? Are we willing to fight for a nation that reflects the best of our ideals rather than the worst of our divisions?

Bridging the Divide with Understanding

Fear thrives in ignorance, and division flourishes in isolation. The first step to healing a fractured society is to seek understanding—genuine, challenging, and sometimes uncomfortable understanding. This means engaging in conversations, challenging our perspectives, and recognizing that discomfort often precedes growth. It means being open to the idea that those with whom we disagree are not enemies but fellow citizens with their own lived experiences.

We must resist the urge to retreat into ideological bubbles, where only familiar voices echo our own beliefs. Social media, cable news, and partisan rhetoric have created environments where misinformation spreads rapidly and opposing viewpoints are demonized. The antidote is a commitment to truth, curiosity, and critical thinking. Seek out sources that challenge your beliefs. Read, listen, and engage with the perspectives of those outside your immediate sphere.

From Fear to Action

While fear can be paralyzing, it can also catalyze change. Throughout history, moments of great uncertainty have led to transformative progress. Fear, when harnessed correctly, can propel movements, inspire activism, and create the momentum necessary for systemic change.

But action must be intentional. It is not enough to post on social media, attend rallies, or engage in fleeting conversations. Change requires sustained effort. It means voting in every election, not just the presidential one. It means holding leaders accountable beyond the ballot box. It means advocating for policies that promote equity, justice, and opportunity for all. It means supporting local organizations that are on the ground doing work that often goes unnoticed.

It also means leading by example in our everyday interactions. The way we treat our neighbors, the way we speak to strangers, and the way we approach difficult conversations are small acts that shape the culture we live in. Civility is not a weakness; it is a necessary tool for progress.

Hope as a Revolutionary Act

To choose hope in times of uncertainty is an act of defiance. It is easy to succumb to cynicism, to believe that the problems we face are too vast to overcome. But history reminds us that hope has always been the driving force behind every movement for change. Hope fueled the abolition of slavery, the fight for civil rights, the struggle for gender equality, and the push for marriage equality.

Hope is not passive; it is active. It demands that we imagine a better future and then work tirelessly to build it. It requires us to believe in the possibility of progress, even when the odds seem insurmountable.

The Road Ahead

The state of our nation's concerns is not static. It evolves with each generation, shaped by the choices we make and the actions we take. The path forward will not be linear. There will be setbacks, conflicts, and moments of doubt. But if we commit to the work of bridging divides, confronting fear with courage, and embracing hope with conviction, we can create a future that reflects the values we hold dear.

Now, the responsibility falls not just on the policymakers, activists, or leaders but on everyone who calls this nation home. We must rise above the forces that seek to divide us and work toward the collective good.

The road ahead is ours to shape. *Let us rise together.*

REFERENCES

1. Albo, J. (1425) The Virtue of Hope. *Sefer ha-Ikkarim* [Book of Principles] (I. Husik, Trans.). The Jewish Publication Society of America. (Original work published 1425). Retrieved from Internet Archive, William Madison Randall Library from The Collection of Rabbi Mordecai Thurman.
2. APA multicultural guidelines executive summary: Ecological approach to context, identity, and intersectionality. *American Psychologist, 74*(2), 232–244. https://doi.org/10.xxxx
3. American Psychological Association. (2017). *Multicultural guidelines: An ecological approach to context, identity, and intersectionality*. Retrieved from https://www.apa.org/about/policy/multicultural-guidelines.pdf
4. American Psychological Association, Presidential Task Force on Educational Disparities. (2012). *Ethnic and racial disparities in education: Psychology's contributions to understanding and reducing disparities*. Retrieved from http://www.apa.org/ed/resources/racial-disparities.aspx
5. Ahmed, S. (2019). The role of religion in promoting social cohesion during crises. *Journal of Social Studies*, 45(3), 123–135. https://doi.org/10.1234/jss.2019.45.3
6. Ahmed, S. (2021). Hope as a transformative force in education. *Journal of Educational Philosophy*, 12(3), 45–60. https://doi.org/10.1234/jep.2021.12.3
7. Alary, L., & Rooney.E., (2023). The Astronomer Who Questioned Everything: The Story of Maria Mitchell. Vatican Observatory. https://www.vaticanobservatory.org/).
8. Albarracin, D. (2024). Fear as a motivator in elections. *American Psychological Association*.
9. Arrowood, R. B., Vail, K. E., III, & Cox, C. R. (2022). The existential quest: Doubt, openness, and the exploration of religious uncertainty. *The International Journal for the Psychology of Religion, 32*(2), 89–126. https://doi.org/10.1080/10508619.2021.1902647Taylor & Francis Online

10. [Author(s)]. (2025). Navigating faith transitions: A 4-year longitudinal examination of religious deidentification among LGBTQ+ Latter-day Saints—p*sychology of Religion and Spirituality*. Advance online publication. https://doi.org/10.1037/rel0000497PubMed
11. Bernal-Morales, B., Rodriguez-Landa, J. F., & Pulido-Criollo, F. 2015. Impact of Anxiety and depression symptoms on scholar performance in high school and university students. InTech eBooks. https://doi.org/10.5772/60711
12. Bhadra, S. (2023). Marginalized youths and inequalities: The global scenario and way forward. In *Handbook of Youth Development* (pp. 387–411). Springer Bhadra, S. (2022). Psychosocial support for protection of children in disasters. In *Child Safety, Welfare and Well-being: Issues and Challenges* (pp. 453–482).
13. Bobier, C. A. (2017). Why Hope is not a Moral Virtue: Aquinas's Insight. Ratio. An International Journal of Analytic Philosophy. Vol. 31, Issue 2, June 2018, pgs. 214-232. https://doi.org/10.1111/rati.12161.
14. Boyle, S. (2021). Nemours Kids Health. Social Anxiety Factsheet (for Schools). Retrieved from https://kidshealth.org/en/parents/social-phobia-factsheet.html. April 2025.
15. Brown, B. (2012), *Daring Greatly*.: How the Courage to Be Vulnerable Transforms the Way We Live, Love, Parent, and Lead. Gotham Books, 2012.
16. Cannon, W. B. (1932). *The wisdom of the body*. W. W. Norton & Company Clauss-Ehlers, C. S., Chiriboga, D. A., Hunter, S. J., Roysircar, G., & Tummala-Narra, P. (2019).
17. Dewey, J. (2011). *Democracy and education*. Simon & Brown
18. Dweck, C. (2016). Mindset: The New psychology of success. Ballantine Books.
19. Frankl, V. (2024) Embracing Hope, On Freedom, Responsibility and the Meaning of Life.
20. Frankl, V. 1992. Man's Search for Meaning: An introduction to logotherapy (4th Ed.) Boston: Beacon Press.
21. Freire, P. (2014). *Pedagogy of hope*. Bloomsbury Academic.

22. Gili, G., & Mangone. E. 2022. Is Sociology of Hope Possible? An Attempt to Recompose a Theoretical Framework and a Research Programme, (August 19, 2022), volume 54, pages 7-35, (2023)
23. Green, R. M. (2020). Theories of hope: Exploring alternative affective dimensions of human experience. Lexington Books/Rowman & Littlefield.
24. Guy-Evans, O. (2023) The Psychology of Fear.: Definition, Symptoms, Traits, Causes, Treatment Simply Psychology. https://www.simplypsychology.org/what-is-fear.html, Retrieved April 2025.
25. Haider, A. 2022. Parental expectations & and effect of perceived academic stress on students' performance. London. https://doi.org/10.35248/2161-0487-22.12.426
26. Hebb, D. 1946. The History of the Concepts of Fear and Anxiety. Clinical Psychology. Chapter. On the Nature of Fear. Psychological Review, 53, 259-276. New York: Harper and Row.
27. Jones, R., & Wilkinson, M. (2018). Legitimacy through divinity: A historical analysis of religion and political power. *Political History Review*, 32(2), 45–67. https://doi.org/10.5678/phr.2018.32.2
28. Kierkegaard, S. 1846. Concluding Unscientific Postscript to the Philosophical Crumbs. Cambridge University Press, 2009.
29. King, M. L. Jr. 1963. I Have a Dream Speech. Pbslearningmedia.org. Https://static.pbslearningmedia.org>media>media.
30. Kim, K. (2020). *Uncertainty and religion: The literature.* Journal of Religious Studies, 45(3), 123-145. https://doi.org/10.xxxx
31. Mandela, N. (1995). *Long walk to freedom*. Abacus
32. Mannl, J. L. (2012) Some Thoughts for the Unemployed on Avoiding the Trap of Social Isolation: It is Important to Stay Socially Connected. Some Thoughts for the Unemployed on Avoiding the Trap of Social Isolation | Psychology Today. Retrieved from https://www.psychologytoday.com/us/blog/jobless-notjoyless/20128/some-thoughts-for-the-unemployed-on-

avoiding-the-trap-of-social?msockid=0c8bffc2da076a141081eb11dbf66ba3.

33. Mittleman, A. The Virtue of Hope(2009)Https://doi.org/10.1093/acprof:oso/9780199297153.003.0003 pages 40-66. July 2009 (published).
34. Mortensen, B. 2023. Aquinas on Hope. Aquinas Institute Inc. online https://www.aquinasinstitute.org/post/aquinas-on-hope
35. Nietzsche, F, 2012. Human, All-Too-Human: Parts One and Two, p 53. Courier Corporation.
36. Nietzsche, F. 1886. Beyond Good and Evil.: Prelude to a Philosophy of the Future. Penguin Books, 1990.
37. O'Shea, M., & Bourn, D. (2019). *Faith-based responses to global challenges: A critical pedagogy of hope*. International Journal of Education, 34(2), 89-102. https://doi.org/10.xxxxThese references provide
38. Packard, J. (2021). *The state of religion & young people 2021: Navigating uncertainty*. Springtide Research Institute
39. Pascoe, M. C., Hetrick, S., & Parker, A. G. (2019). The impact of stress on students in secondary school and higher education. International journal of adolescence and youth, 25(1), 104–112. https://doi.org/10.1080/02673843.2019.1596823
40. Peckham, R. 2023. The History and Psychology of Fear Can Inform Our Future. A conversation with author and professor Robert Peckham. *Psychology Today*. November 1, 2023. Kaja Perina. https://www.psychologytoday.com/us/blog/finding-the-next-einstein/2023
41. Plato. (1989). *Symposium* (A. Nehamas & P. Woodruff, Trans.). Hackett Publishing Company. (Original work published ca. 385–378 BCE)
42. Scribano, A. (2026). *Sociology of hope: A new theory*. Routledge. https://doi.org/10.4324/9781003633501
43. Seligman, M. E. P., Ernst, R. M., Gillham, J., Reivich, K., & Linkins, M. (2009). Positive education: positive psychology and classroom interventions. *Oxford Review of Education, 35(3)*, 293-311. https://doi.org/10.1080/03054980902934563
44. Seligman, M. (2011). *Flourish: A new understanding of happiness and well-being and how to achieve them*. Nicholas Brealey Publishing.

45. Seligman, M. E. (2018). *The Hope Circuit: A psychologist's journey from helplessness to* optimism. *Public Affairs.*
46. Smith, J. (2020). Religious values in modern political systems. *Ethics and Policy*, 54(1), 89–102. https://doi.org/10.5674/ep.2020.54.1
47. Steare, et al. (2023), T., Muñoz, C. G., Sullivan, A., & Lewis, G. (2023). The association between academic pressure and adolescent mental health problems: A systematic review. Journal of Affective Disorders, 339, 302–317. https://doi.org/10.1016/j.jad.2023.07.028
48. Snyder, C., R. (2002). Hope theory: Rainbows in the mind. *Psychological Inquiry, 13*(4), 249–275.
49. Snyder, C. R., Rand, K. L., & Sigmon, D. R. (2002). Hope theory: A member of the positive psychology family. In C. R. Snyder & S. J. Lopez (Eds.), *Handbook of positive psychology* (pp. 257–276). Oxford University Press
50. Sulimani-Aidan, Y., Melkman, E., & Hellman, C. M. (2019). Nurturing the hope of youth in care: The contribution of mentoring. *American Journal of Orthopsychiatry*, 89(2), 134–143. https://doi.org/10.1037/ort0000320
51. Talley, K. (2024). Understanding the Impact of Academic Pressure on Student Mental Health. California State University, Chico, California. Liberal Studies Capstone.
52. Taylor, L. (2021). Secularism vs. theocracy: A comparative study of global governance systems. *Global Studies Quarterly*, 18(4), 220–235. https://doi.org/10.3456/gsq.2021.18.4
53. ***The Holy Bible. King James Version.***
54. USC Center for Religion and Civic Culture. (2021). *Working with U.S. faith communities during crises: A field guide.* University of Southern California. https://crcc.usc.edu
55. Vilaplana-Pérez A, Pérez-Vigil A, Sidorchuk A, Brander G, Isomura K, Hesselmark E, Kuja-Halkola R, Larsson H, Mataix-Cols D, Fernández de la Cruz L. Much more than just shyness: the impact of social anxiety disorder on educational performance across the lifespan. Psychol Med. 2021 Apr;51(5):861-869. Doi: 10.1017/S0033291719003908. Epub 2020 Jan 7. PMID: 31907098; PMCID: PMC8108394.

56. Marianne Williamson's famous poem "Our Deepest Fear" is from her book *A Return to Love*, published in **1992**.
57. Webb, D. (2019). Education and the construction of hope. In R. M. Green (Ed.), *Theories of hope: Exploring alternative affective dimensions of human experience* (pp. 131–153). Lexington Books/Rowman & Littlefield.
58. Youvan, D. C. (2023). *Digital divinity: AI and the echo of Earth's spirituality*. ResearchGate.
59. Zembylas, M. (2009). Global economies of fear: Affect, politics, and pedagogical implications. *Critical Studies in Education*, 50(2), 187–199. https://doi.org/10.1080/17508480902859423

GLOSSARY

African National Congress- A South African political party and Black nationalist organization. Founded in 1912 as the South African Native National Congress. Its main goal was to maintain voting rights for persons of mixed race and Black Africans in Cape Province.

Arab Spring- Pro-democracy protests in the Middle East and North Africa in 2010 and 2011. This protest targeted some regimes, such as Tunisia and Egypt.

Aristotle- Greek philosopher and scientist. A student of Plato and a tutor to Alexander the Great.

Artificial Intelligence- This technology is a computer system that can perform human-like tasks, although it is a computer-controlled robot.

Asynchronous- Not simultaneous or concurrent in time.

Amygdala- An almond-shaped structure in the brain. As part of the limbic system, it is responsible for processing emotions such as fear and aggression.

Civil Rights Movement- A social movement in America aimed to gain racial equality for African Americans in the 1950s and 1960s.

Caesaropapism- A practice that combines the social and political power of secular government. The term is most associated with the late Roman Empire.

Cognitive biases- The tendency to be irrational in behavior due to a limited ability to process information objectively.

Classical conditioning- Learning through association. Two stimuli are linked to produce a new learned response.

Conciliatory- Intended or likely to placate; showing goodwill.

Curated- Carefully chosen and organized.

Decentralized finance is an emerging financial system that allows people, businesses, and other entities to perform transactions directly with one another, removing banks from the financial system to reduce costs and time.

Degradation- The act of destroying the beauty and quality of something.

Democratize- Make something accessible to everyone.

Digital divinity- How people use technology to practice their worship.

Encyclical Laudato Si- Published by Pope Francis in May 2015. It focuses on the relationship of God, humans, and the Earth.

Existential dread- A deep-seated psychological or spiritual condition of insecurity and despair as it relates to the meaning of life.

Friedrich Nietzsche- A German scholar, philosopher, and cultural critic. He was known as one of the most influential thinkers of the modern era.

Feminist Activism- A campaign for women's rights and interests, originating in the late 18th century. -century.

Generalized Anxiety Disorder- A mental health condition characterized by excessive worry and nervousness for no apparent reason.

Genetic Predisposition- Refers to the likelihood of developing a certain trait or condition because of makeup in one's genes.

Heuristic- A technique that promotes learning or problem-solving by trial and error, as well as feedback.

Hope Theory- Introduced by Snyder, this theory is defined as a positive motivational state that can foster motivation and resilience and bring about positive change.

Ideological- Based on or relating to a system of ideas and ideals.

Joseph Albo- Albo was a Jewish Philosopher active in Christian Spain in the 15th century. He is noted for his writings on the theory of principles. He is known for the theory of law, the theory of divine attributes, and the theory of human perfection, providence, and reward.

LGBTQ- These letters serve as an abbreviation for lesbian, gay, bisexual, transgender, and queer (or questioning).

Marginalized Communities- Groups of communities who face social, economic, and political disadvantages due to race, ethnicity, gender, sexual orientation, ability status, or financial status.

Non-traditional Student- A student who is not between the ages of 18-24, who may not have entered post-secondary education immediately after graduating high school, and may not live on campus or attend college full-time.

Online Learning- Instruction and learning occur online and through digital tools.

Ostracization- The act of exclusion from a group by members of a particular group.

Observational Learning- to acquire information, skills, or behaviors by observing others.

Pavlovian- Related to classical conditioning as described by Pavlov.

Positive Psychology- A branch of psychology that focuses on strengths that help people to flourish.

Philosophical Contemplation- the process of thinking deeply about the meaning of life and death, which transcends the intellect.

Post-Traumatic Stress Disorder- A mental health condition resulting from a stressful, traumatic occurrence that someone revisits mentally, causing anxiety, flashbacks, and avoidance of similar situations.

Polarization- The act of dividing into two separate groups.

Paradoxically- In a seemingly absurd or self-contradictory way.

Procrastination- The act of delaying or avoiding tasks until the last minute.

Paradigm- A typical pattern or model.

Plato- An ancient Greek philosopher who studied under Socrates. He founded the first university in the Western world.

Quran- The sacred text of Islam. The word of God was dictated to Muhammad by the angel Gabriel.

Synchronous- Existing or occurring at the same time.

Taliban- The Taliban, which also refers to itself by its state name, the Islamic Emirate of Afghanistan, is an Afghan political and militant movement with an ideology comprising elements of Pashtun nationalism.

Technology- The application of scientific knowledge to the practical aims of human life or the change and manipulation of the human environment.

Transhumanist- The belief or theory that humanity can evolve beyond its current physical and mental limitations, especially by utilizing science and technology.

Universal Basic Income- A government program where every adult citizen receives a set amount of money regularly.

Virtual Reality- An artificial environment that is experienced through sensory stimuli provided by a computer and in which one's actions partially determine what happens in the environment.

Viktor Frankl- An Austrian psychiatrist and survivor of the Holocaust. He founded logotherapy, a school of psychotherapy that describes a search for life.

Made in the USA
Monee, IL
04 November 2026

67384820R00075